EXTINCTION

THE FUTURE OF HUMANITY

A SHORT STUDY OF EVOLUTION
FROM THE ORIGIN OF LIFE TO THE PRESENT

RONALD E. SEAVOY

ISBN 978-0-88839-691-4

Cataloging in Publication Data

Seavoy, Ronald E.
Extinction : the future of humanity : a short study of evolution from the origin of life to the present / Ronald E. Seavoy.

Includes bibliographical references and index.
Also available in electronic format.
ISBN 978-0-88839-691-4

1. Life—Origin. 2. Evolution—Religious aspect—Christianity. 3. Extinction (Biology). I. Title.

QH366.2.S44 2009 576.8 C2009-904755-1

Printed in South Korea — PACOM

Maps: Daniel Henry Schellhas

Published simultaneously in Canada and the United States by

HANCOCK HOUSE PUBLISHERS LTD.
19313 Zero Avenue, Surrey, BC Canada V3S 9R9
(604) 538-1114 Fax (604) 538-2262

HANCOCK HOUSE PUBLISHERS
1431 Harrison Avenue, Blaine, WA, USA 98230-5005
(604) 538-1114 Fax (604) 538-2262

Website: **www.hancockhouse.com**
Email: **sales@hancockhouse.com**

If you are of religious temperament and want certitude for your faith, this book is not for you. To enjoy this book there must be a small voice in your psyche that is willing to pry up the lid of faith in order to see a few rays of scientific light. It is hoped that this light will help you understand that the real foundation of religion is moral teachings that exclude myth, magic and miracles.

CONTENTS

Eras of Geologic Time

MILLIONS OF YEARS	GEOLOGIC ERAS
Present	**mass extinction**
65–present	Tertiary (Cenozoic)
65	**mass extinction**
142–65	Cretaceous
206–142	Jurassic
206	**mass extinction**
251–206	Triassic
251	**mass extinction**
290–251	Permian
362–290	Carboniferous (Mississippian, Pennsylvanian)
362	**mass extinction**
417–362	Devonian
438–417	Silurian
435	Plant and animals colonize land
438	**mass extinction**
495–438	Ordovician
495	**mass extinction**
550–495	Cambrian
610–550	Vendian
555–550	Ediacaran Assemblage
610	First metazoan animals
700	First metazoan organisms
1500	First eukaria cells
3500	First fossil bacteria
4000	Approximate beginning of life
4600	Beginning of planet earth

PREFACE

This book examines humans as another animal in the biosphere that is temporarily at the top of the food chain. Like other animals in the biosphere, humans are subject to the same stresses of nature that have caused the extinction of multi-millions (or billions) of species from the appearance of life to the present. The geologic record is clear. Larger and more complex organisms are the first animals to become extinct during stressful events in the biosphere. There is nothing inevitable about humans evolving on earth or humans surviving to the end of the solar system. Humans are vulnerable to extinction in similar ways that the fossil record documents the extinction of all previous animals at the top of the food chain.

CHAPTER 1

APPEARANCE OF LIFE

> Then God said: Let us make man in our image, after our likeness; and let man have dominion over the fish of the sea, and over the birds of the air, and over the cattle, and over all the earth.... So God created man in his own image, in the image of God he created him.... And God blessed humans and God said to them: Be fruitful and multiply, and fill the earth and subdue it; and have dominion over the fish of the sea and over the birds of the air and over every living thing that moves upon the earth. — Genesis 1: 26–28

Life is the ability of organisms to self-replicate. Life began when prebiotic molecules were encapsulated long enough in a prebiotic cell for some molecules to self-assemble (synthesize) into larger molecules required for self-replication. Prebiotic cells and prebiotic molecules are inorganic chemical compounds that form in favorable environments, often in very large amounts. Prebiotic cells became living organisms when they acquired the ability to sustain self-replication. Could self-replication have happened because a great spirit (god) caused it to happen? Unlikely. Could self-replication have happened because of spontaneous chemical reactions? Probably.

The god explanation for the origin of self-replicating organisms is based on revelation. In the Judeo-Christian-Muslim religions the truths of revelation are validated by deductive reasoning. An omnipotent being revealed his purpose for creating life to humans and then directed the scope of life for human benefit.

All the complexities of the biosphere derive from god's purpose of making humans the governor of the biosphere. If a person believes that life is the product of an omnipotent god, then a person can deduce all biological complexities from the power and purpose of this god. Immediately or ultimately geological and biological sciences are superfluous. If a person believes the Genesis myth is fact, all biological sciences not directly related to medicine or agriculture are marginal appendages to true belief.

If a person believes that the biosphere is a product of intelligent design because god created it for the benefit of humans, the principal purpose of scientific inquiry is to confirm the validity of the Genesis myth. This myth asserts that god created all life for the benefit of humankind. Humans are the final biotic creation in his design of the universe because he created humans in his image. Humankind is a reflection of god's omnipotence; therefore, humans must glorify him. Deductive reasoning from the Genesis myth gives purpose to human life in terms that are understandable by persons who are scientifically illiterate.

In its most bizarre form, large numbers of Christians reject the evolution of life from bacteria cells to chordate worms, to fish, to terrestrial vertebrates, to great apes, to humans. They favor a highly simplistic explanation. They believe that, in the recent past, god created humankind and all other organisms in the exact anatomical form they now exist. They are assured this is true by a literal interpretation of infallible scripture that assigns humankind dominion over all forms of life. They believe this even though the Genesis version of creation is one of many mythical accounts of creation composed by preliterate societies to explain how, why and where they came to occupy the geographic region where they live.

Religious faith is the only basis for claiming that god created self-replicating cells; or that god designed self-replicating cells to evolve into humankind; or that god created humans in the exact anatomical form they now exist. Religious faith allows clergymen to use deductive reasoning to validate the claim that humankind is a special creation by god to govern the earth's biosphere. The primacy of deductive reasoning allows clergymen to deny the existence of evolution in order to elevate humankind into god's surrogate on earth. Deductive reasoning (based on revelation) allows scientific explanations to be ignored or denied.

The scientific explanation for the origin of self-replicating cells is based on inductive reasoning. Inductive reasoning requires experiments that can be confirmed by replication. These experiments are a foundation for additional experiments that increase the scope of the original experiment. Inductive reasoning is used to organize large amounts of data into coherent explanations that have a high degree of certainty.

Life came into existence because a series of chemical reactions within one (or more) prebiotic cells made it possible. Many of these chemical reactions can be duplicated by laboratory experiments. The inductive reasoning of science gives no purpose for the origin of self-replicating cells or their evolution into complex organisms. *How* life came into existence is the question asked by science because this question can be tested by experiments. *Why* life came into existence is not a question asked by scientists because there is no way to test this question.

PRECONDITIONS

What were the conditions that made life possible? This is an open question, but four physical conditions were essential for life to begin: (1) little or no oxygen in the atmosphere because oxygen would rapidly degrade prebiotic molecules to carbon dioxide and water; (2) warm oceans that had many favorable microenvironments where prebiotic cells and molecules could concentrate; (3) prebiotic cells that could concentrate favorable mixtures of prebiotic molecules; (4) a source of energy that could be captured by prebiotic molecules in order to assemble other prebiotic molecules into self-replicating chemicals. Sunlight, ultraviolet radiation, lightning and submarine volcanic vents (black smokers) were instantly available sources of energy.

Scientific researchers are trying to duplicate the origin of prebiotic cells and then map how encapsulated prebiotic molecules were synthesized into molecules that had the ability to self-replicate. Their research is guided by the hypothesis that sometime about four billion years ago there were favorable conditions somewhere in the oceans where simple prebiotic molecules were encapsulated by prebiotic cells, and that these cells endured long enough for encapsulated chemical compounds to synthesize into molecules that could self-replicate. In other words, the hypothesis that life had a spontaneous chemical origin allows the origin of life to be investigated by scientific experiments.

Natural concentrations of prebiotic molecules have been found in some contemporary marine microenvironments. These conditions were almost certainly present in favorable marine microenvironments when life first became possible. Sustainable replication cannot occur in random mixtures of dispersed molecules. Sustaining life requires that replication takes place in cells that concentrate essential molecules in order for them to synthesize into the larger molecules essential for life. Prebiotic cells are formed by membranes that have the ability to encapsulate the molecules necessary for self-replication. The cells must endure long enough for the molecular chemistry of replication to occur. How long is long? One second, a minute, five minutes, ten minutes, an hour, several days?

The working hypothesis of evolutionary biologists is that several chemically active prebiotic molecules could react to form new chemical compounds if they were confined in a cell. A prebiotic cell had to exist long enough for some encapsulated molecules to react with other encapsulated molecules to synthesize molecules that could self-replicate. At the origin of life, prebiotic chemical reactions within prebiotic cells became biological when nonliving chemicals within a prebiotic cell synthesized themselves into chemical compounds that could sustain self-replication. The self-replicating cell became an organism. Life began.

Self-replication required an ability to capture energy and acquire nutrients in order to replicate. Sunlight was the most likely source of energy, meaning the first self-replicating cells were probably photosynthetic (organic compounds formed through the addition of light energy). These cells encapsulated prebiotic molecules that could catalyze sugars from water and carbon dioxide. The other possibility is prebiotic cells using the thick soup of mineral nutrients at geothermal vents to synthesize self-replicating molecules. As of now, however, no prebiotic cells have been found in this environment.

Sugars synthesized by the first cells were the nutrients that sustained replication. Self-replication may have happened several times immediately after the earth became habitable but only one pathway has survived. This is the pathway used by photosynthetic molecules and its variant, chemosynthesis (during which organic compounds are formed by energy derived from inorganic chemical compounds), used by molecules that inhabit geothermal vents. This pathway is the basis for all subsequent life from the most primitive archea to humans. After self-replication became a self-sustaining chemical reaction (called life), life had only one purpose: reproduction.

SELF-REPLICATING CELLS

Sustaining life required that the chemistry of replication take place in a cell. A physical mechanism that produces prebiotic cells has been discovered. The mechanism is a cycle of drying and wetting of some naturally occurring prebiotic lipids. Lipid molecules self-assemble into prebiotic cells that have two layers. All cells of all organisms have walls composed of two layers. In several contemporary marine microenvironments prebiotic lipids concentrate in a broth.

One of the most favorable microenvironments is intertidal zones in desert climates where stranded liposome layers dry during low tide and rehydrate during high tide. Rehydrated lipid membranes self-assemble into two-layer membranes in the shape of spheres and cylinders (rods). These are liposome cells. Laboratory experiments have observed prebiotic liposome cells encapsulate many prebiotic molecules.

Liposome cells are stable as long as they remain in water. This means that there is time for molecular chemical reactions to occur within liposome cells. It is highly probable that encapsulation of chemically active prebiotic molecules contributed to the assembly/synthesis/polymerization of simple proteins in liposome cells.

Something similar has recently been discovered. Fullerenes (buckyballs) are a newly discovered crystalline form of carbon that is now known to be very common in the contemporary environment. Buckyballs are inorganic cells. They have hollow interiors that can encapsulate a large variety of molecules. Buckyballs have been discovered in sediments that are more than 250 million years old. And searching for them in older sediments will probably find them because they are very stable at atmospheric or near atmospheric conditions. This is evidence that prebiotic cells can exist in large numbers over extended periods of time.

Almost certainly, favorable marine microenvironments in the Archean era (4.0 to 2.2 billion years ago) contained a large variety of prebiotic organic molecules that were available for encapsulation in prebiotic cells. Experiments conducted by molecular chemists in an atmosphere without oxygen have produced large quantities of prebiotic organic molecules by physical processes.

The gases used to create preorganic molecules in laboratories are water vapor, carbon dioxide, methane and nitrogen. They are confined in a large glass vessel and electricity (lightning) and ultraviolet radiation is passed through them. A large number of sugars, amino acids, short strands of RNA (ribonucleic acid) and other simple molecules have been synthesized. This is strong evidence that prebiotic organic molecules were synthesized in large quantities from these same gases that composed the earth's early atmosphere. These prebiotic molecules were concentrated in a prebiotic broth in favorable marine microenvironments. There they were encapsulated by prebiotic cells.

Almost certainly, from the inception of life until about 2.2 billion years ago, the earth's atmosphere and marine water had minimum amounts of oxygen. If there had been an oxygen-rich atmosphere, as exists today, prebiotic organic molecules would have been reduced to water and carbon dioxide almost as soon as they formed. Absence of oxygen in the atmosphere and marine water meant there was no ozone layer. Ultraviolet radiation reached the earth's surface and provided energy for synthesizing prebiotic molecules. Energy also came from lightning and visible light.

When did life originate? Almost certainly it occurred early in the Archean era, about four billion years ago. It began very soon after the earth became habitable, when warm water oceans covered most of its surface. Radiometric dating of meteorites and moon rocks indicates that the solar system—and earth—were formed about 4.6 billion years ago. The oldest rocks preserved in the earth's geologic record are about 3.8 billion years old, and the first evidence of life are microfossils of archea (primitive bacteria). They have been discovered in early Archean sediments that are about 3.5 billion years old. These sediments are found in Australia and South Africa.

A principal means of producing prebiotic molecules for encapsulation in prebiotic liposome cells was synthesis on templates of crystalline minerals. At least three possible templates exist in contemporary marine microenvironments (and also existed in the early Archean era). They are: (1) clay crystals with a high copper content held in suspension by winds, (2) surfaces of pyrite and other detrital crystals in sea floor sediments, (3) bubbly froth created by wind on water in bays that contained concentrations of prebiotic molecules.

Within liposome cells, prebiotic amino acids and other simple molecules were catalyzed and polymerized into the larger molecules necessary for life to begin. In the trillion, times a trillion, times a trillion, times a trillion, times a trillion that short-lived prebiotic cells formed every day in favorable marine microenvironments, there was a chance that one cell would capture the right mixture of prebiotic amino acids and other prebiotic molecules needed for prebiotic RNA to synthesize the correct mixture of molecules that could initiate and sustain self-replication.

Self-replication of cells occurred when several prebiotic molecules attached themselves to prebiotic RNA encapsulated in prebiotic cells. RNA formed a scaffold within the cell for synthesis to occur. Synthesis produced the pigment molecules that synthesized sugars using light as energy and the protein molecules needed to sustain cell replication.

Where did the concentrations of prebiotic molecules needed for synthesis of life come from? Almost certainly they came from prebiotic cells that disintegrated in storms, but which had existed long enough for some synthesis to take place within them. The prebiotic sugars, amino acids and other organic molecules from disintegrated cells were then available for encapsulation in existing or new prebiotic cells. In favorable marine microenvironments, there was no shortage of prebiotic molecules available for encapsulation in prebiotic cells.

The two places where life (archea) could have originated are: (1) warm water in tidal pools, (2) hot water around submarine volcanic vents. Energy and mineral nutrients were super abundant around submarine volcanic vents. If archea originated there, they acquired energy by breaking the chemical bond of hydrogen sulfide. This energy was used to break the bond of carbon dioxide and water in order to synthesize sugars. This is called chemolithic synthesis. Chemolithic metabolism is a highly efficient source of energy to sustain replication.

In the contemporary biosphere a high percentage of archea live in extreme anaerobic microenvironments. The most accessible of these environments is the peripheral water of volcanic springs that vent boiling water, but they also thrive around submarine volcanic vents (black smokers) found in midocean rift zones. Four billion years ago there was an incredibly large amount of submarine volcanism. Hydrothermal vents (black smokers) that are associated with this volcanism, vented huge volumes of hydrogen sulfide, iron sulfide and carbon dioxide gas into ocean water. Submarine volcanic vents had abundant energy and mineral nutrients. Although energy and mineral nutrients were ample, experiments have not yet discovered prebiotic cells.

Alternatively, if archea originated in warm tidal pools, they could sustain metabolism by using sunlight to break the chemical bonds of carbon dioxide and water to synthesize sugars. This is photosynthesis. Photosynthesis and chemolithic synthesis supplied cells with an internal source of food by converting solar energy and chemolithic energy into sugars that sustained metabolism. For more than 1.5 billion years, life concentrated and evolved in one or both of these niches.

Laboratory experiments have produced prebiotic RNA molecules. RNA molecules alone can synthesize some proteins that store sufficient sequencing information to form a replicating pathway for the entire cell. These proteins are called enzymes. Enzymes control replication and metabolism by controlling the speed, sequence, function and location of all protein cells in all organisms. They are the biologic engines of cell replication.

Replication by RNA alone was inefficient because the RNA pathway makes many mistakes. (RNA has no repair mechanism but DNA does.) Nonetheless, laboratory experiments indicate RNA was sufficiently efficient to initiate and sustain cell replication. Some of the mistakes in replication were beneficial because after the first appearance of life, mutations in the RNA pathway began the evolution into the DNA pathway. This indicates that natural selection was contemporary with the appearance of life.

Efficiency and speed of replication were improved by the attachment of more proteins on longer strands of RNA within cells. This was first step in the synthesis of DNA that vastly increased the efficiency of replication. Efficiency of replication by DNA rapidly displaced cells that replicated by RNA alone. No organisms using RNA alone have been discovered in the contemporary biosphere. The great unanswered question of molecular and evolutionary biology is how RNA evolved into DNA. The earliest microfossils are archea. The morphology of these cells closely resembles the morphology of archea that live in deep marine mud and other extreme environments.

Chemolithic and photosynthetic metabolism frees two hydrogen molecules. In chemolithic metabolism the sulfur molecule from the metabolism of hydrogen sulfide is added to FeS (pyrrhotite), a mineral vented in huge quantities by black smokers. This reaction synthesizes pyrite (FeS2). Pyrite is the waste product of chemolithic metabolism. There are huge quantities of pyrite around volcanic vents from the Archean era to the present. The two oxygen molecules that are the waste product of photosynthesis reacted with soluble iron to form magnetite or soluble calcium to form limestone.

During the Archean era, chemolithic archea/bacteria deposited huge quantities of carbon and pyrite in deep-water black shale. The black was caused by immense quantities of dead microbes incorporated into sediments. Chemolithic microbes were hugely abundant in Archean oceans because there was an abundance of mineral nutrients and there were no animal predators.

From my experience as a geologist exploring for mineral deposits on the Canadian Shield, I know there are incredibly large amounts of carbon in volcanics and black sediments of Archean age (3 to 2.2 billion years ago). In most places the carbon is now graphite. There are many thousands of lenses of graphite and massive pyrite sandwiched between lava flows. Frequently the graphite lenses are fifty meters thick. In the Archean era, chemolithic microbes prodigiously multiplied wherever there was submarine volcanism. Originally, the graphite was a trillion, times a trillion, times a trillion, times a trillion, times a trillion, times a trillion chemolithic archea/bacteria cells that accumulated in depressions around hydrothermal vents associated with submarine lava flows

In most places, metamorphism has obliterated all evidence of the microbial origin of graphite, but the biogenic origin of graphite is not in doubt. These accumulations could not have occurred unless there were huge abundances of hydrogen sulfide and carbon dioxide gases venting from submarine volcanism and equally huge amounts of microbial life—as exists around many contemporary black smokers found along midocean rift zones.

In my opinion, the best location for the origin of life is in warm tidal pools or shallow marine bays where evaporation concentrated a broth of prebiotic molecules. This environment also produced an abundance of prebiotic liposome cells; additionally, solar and ultraviolet energy was always available as a source of energy. But the spontaneous origin of life was equally possible by chemosynthesis. Although photosynthetic and chemolithic microbes appear to be genetically different, they share the same RNA–DNA pathway of cell replication except they use a different source of energy to sustain the metabolism that sustains reproduction.

Sometime around three billion years ago one specie of archea evolved into bacteria. Bacteria were more efficient synthesizers of sugars and other organic molecules used for the metabolism that sustains reproduction. All of the earliest preserved microfossils are in shallow-water sediments, and the best preserved are associated with kerogen.

Kerogen is a hydrocarbon derived from organisms. It contains a higher percentage of carbon-12 isotopes than atmospheric carbon dioxide because microorganisms preferentially use carbon-12 rather than the heavier isotope carbon-13. Kerogen is a sure indicator of biologic activity.

Soon after photosynthetic bacteria evolved, some species became colonial. They formed mats, oncolites and stromatolites in shallow water. Mats and stromatolites prevented storms from dispersing them to less hospitable environments. A colonial habit also facilitated the exchange of genes so that beneficial mutation became more frequent (measured in units of 100 million years).

Red-colored sediments are very rare in the Archean era because both the atmosphere and marine water had minimal amounts of oxygen. There was, however, abundant iron dissolved in marine water. About 2.2 billion years ago red sediments became common in stratigraphic sequences. The first appearance of abundant red-colored rocks measures the changed composition of the atmosphere. The oxygen that changed the composition of earth's atmosphere was the waste product of photosynthetic bacteria.

The red was caused by iron oxide that was concentrated in iron formations that were deposited in shallow water where oxygen was continually renewed by the photosynthetic bacteria.

Iron formations are unambiguous evidence that an oxygen-rich atmosphere came into existence because iron-rich sediments (banded iron formations) of this age are found on all continents. They are the source of 90 percent of the ores that are smelted into metallic iron. Some geologists characterize the precipitation of soluble iron from marine water as the rusting of the oceans. This marks the end of the Archean era.

Oxygen produced by photosynthetic bacteria transformed the earth's atmosphere from little or no oxygen to approximately the 20 percent that exists today. The oxygen came from bacteria that evolved the ability to live in an oxygen-rich atmosphere that they created. Many anaerobic microbes evolved into aerobic microbes or they became extinct. The oxygenation event may be related to the evolution of the RNA pathway of reproduction into the RNA–DNA pathway of reproduction.

Did photosynthetic and chemolithic microbes originate according to the above analysis? Did photosynthetic archea/bacteria evolve into chemolithic microbes or vice-versa? Little evidence is preserved in living microbes of how cell replication began and became self-sustaining or where it first occurred—but it did occur. We know this with certitude; otherwise, life would not be self-sustaining.

Investigating how life originated is expensive and time consuming because it must be done by highly trained persons at generously funded laboratories. Complex instruments must be built, techniques developed to use them, knowledge accumulated and the accumulated knowledge accurately interpreted. Investigations to discover how replication began are in their infancy because molecular biology, and its sister discipline evolutionary biology, did not become separate disciplines until the 1970s.

As of now, there is only limited understanding of how life originated; although, there is considerable understanding of how specific enzymes catalyze the proteins required to sustain life.

We do know, however, that during the first two billion years of life (from about four billion to about two billion years ago), there were many beneficial mutations within archea and bacteria cells that helped improve reproduction and expand the number of niches where life could reproduce.

MOLECULAR BIOLOGY

Molecular biology investigates the chemical reactions that take place in the cells of all organisms. Put in other terms, molecular biology seeks to understand the chemical origin of life and how reproduction is sustained. It focuses on the structure of enzymes and how enzymes are synthesized within cells. Enzymes are organic catalysts that can synthesize themselves and other proteins necessary for cell replication in the RNA–DNA pathway.

The origin of life probably began when a prebiotic cell encapsulated several short strands of prebiotic RNA. RNA is a single-strand polymer that can replicate itself and synthesize some protein molecules that contribute to replication. RNA is not a protein but protein molecules (enzymes) can attach to long strands of RNA, and long strands can store sufficient information for sequencing the molecular chemical reactions necessary for replication.

Laboratory experiments indicate that some common clay crystals can act as templates to synthesize short strands of prebiotic RNA. Inorganic templates are the probable origin of prebiotic RNA. This mode of synthesis could have produced very large numbers of RNA molecules. From the perspective of evolutionary biologists, life began with the ability of short strands of encapsulated RNA molecules to polymerize into longer strands of RNA that could sequence enough information for cells to replicate themselves.

Laboratory experiments confirm that the RNA molecule alone can: (1) encode genetic information, (2) encode sequences of molecular reactions, (3) catalyze molecular reactions that reproduce cells by division (mitosis), (4) catalyze simple proteins. There is rapidly accumulating evidence that pre-archea cells reproduced by the RNA pathway alone.

The first self-replicating cells were pre-archea. These cells retained their integrity long enough for RNA molecules to replicate along a pathway that was transferable to daughter cells by division. They also had the ability to synthesize or encapsulate food. Subsequently, sites on longer strands of RNA attached encapsulated prebiotic amino acids (or simple proteins) that improved: (1) replication, (2) metabolism, (3) strengthened cell membranes, (4) increased the ability to absorb molecules through cell membranes, (5) encoded increasing amounts of genetic information for sequencing the chemical reactions necessary that improved the precision of replication.

There is overwhelming circumstantial evidence that the RNA pathway was the prerequisite for pre-archea cells to evolve into archea cells and then evolve into bacteria cells that use the RNA–DNA pathway of replication. The RNA–DNA pathway of replication is found in all cells of all organisms today. The circumstantial evidence for this means of replication (and evolution) is that no other microorganisms have been discovered that could evolve this pathway of replication.

After the RNA–DNA pathway evolved, replication by the RNA pathway alone became extinct. If pre-archea cells still exist they have not yet been discovered. All organisms now living in the biosphere use the RNA–DNA pathway of reproduction. There are no known exceptions.

Replication, however, could not be sustained unless there was a metabolic source of energy. Chlorophyll produced the needed energy—after prebiotic chlorophyll molecules were encapsulated in a prebiotic cell or, alternatively, chlorophyll was catalyzed from amino acids encapsulated in pre-archea cells.

Chlorophyll is an enzyme that uses the sun's energy (light) to break the chemical bond of water and carbon dioxide in order to synthesize sugars. Life originated at the moment when photosynthesis by chlorophyll or related enzymes became a sustainable chemical reaction within a cell that used an RNA molecule to replicate itself. Current research in molecular chemistry indicates that molecular chemists will soon be able to define most of the chemical reactions necessary for reproduction by RNA alone.

There is little evidence of how the RNA pathway evolved into the RNA–DNA pathway, but it is possible to make an intelligent guess. All living bacteria continually exchange genes that are composed of RNA–DNA molecules. It is probable that the exchange of RNA fragments with attached protein molecules, plus the encapsulation of amino acid molecules, contributed to the evolution of the RNA–DNA pathway.

The great advantage of the RNA–DNA pathway is that replication became highly predictable compared to the molecular chemistry of the RNA pathway. In the RNA pathway lethal mutations (mistakes) were frequent, but the number of mutations increased the possibilities of beneficial mutations.

Beneficial mutations propelled the evolution of the RNA pathway into the RNA–DNA pathway. The chemical reactions in the RNA–DNA pathway are the result of an immense number of natural selections that made it possible for pre-archea cells to evolve into archea and archea to evolve into bacteria. Bacteria could occupy many more marine niches than archea.

Reproduction requires a code (map) that regulates replication and metabolism. In all contemporary organisms the replication code is stored in genes and chromosomes.

What are genes? What are chromosomes? Genes are short

strands of DNA that have attached enzymes that synthesize proteins, as well as encode the sequences of molecular synthesis necessary for reproduction. Like the RNA molecule, the DNA molecule is not composed of proteins. It is a nonprotein scaffold where genes and enzymes are attached in highly predictable sequences. Enzymes synthesize themselves, catalyze the synthesis of other proteins, encode the sequence of protein synthesis and emplace them in the organism. Chromosomes are long strands of genes that contain all of the genetic information required for the reproduction of an organism.

Like RNA, DNA is a long strand (polymer) of small molecules, but unlike RNA it forms a double helix. The double helix has two great advantages compared to RNA. It is the scaffolding for an infinite number of variations and it has a self-repair mechanism that insures almost all replications are exact copies.

The RNA–DNA pathway that evolved in archea probably occurred 3 billion years ago. This pathway has been hyperconserved because it is the core chemistry of reproduction. Its hyperconservation is strong circumstantial evidence that the molecular chemistry that evolved the RNA–DNA pathway happened only once because reproduction of all cells in all existing organisms use it, including humans. Put in another perspective, the genes and chromosomes that govern human reproduction are not unique. They are a variation in the package of genes and chromosomes that govern the reproduction of all organisms.

The DNA molecule has a curious scaffolding. In all humans there are many spaces on it that appear to have no function. They are blank. Blank spaces are about 97 percent of human DNA. It is highly probable that at some time during the one billion years (four to three billion years ago) that it took archea to evolve into bacteria these blank spaces were sites for genes that regulated replication and metabolism.

During this long evolution they were turned off whenever a

mutation increased the efficiency of replication and metabolism. More efficient enzymes were accommodated by adding them to new sites on the DNA scaffold. Older gene sites remained in place but were disabled by becoming noncoding sites. Alternatively, the noncoding (blank) sites perform sequencing or other functions that have not yet been discovered.

The sequencing of molecular chemistry by RNA–DNA is highly predictive but is not perfect. There are numerous mutations. Most mutations are repaired or are benign, but lethal mutations are terminated by natural selection or perhaps some survive as short lengths of DNA after the death of the microbe. This could be the origin of viruses.

VIRUSES

Viruses are short lengths of DNA. Some have only one helix of DNA. The simplest viruses have ten genes or less, compared to several thousand in the cells of the simplest archea and bacteria. Viruses are not alive because they lack the ability to self-replicate; or perhaps they are half-alive. They are composed of short lengths of DNA that use enzymes to pierce the membranes of healthy cells. After they enter a cell, other enzymes activate the cell's enzymes and appropriate newly synthesized proteins for their replication and metabolism. During replication viruses become alive. When replication is complete they crystallize into a giant molecule because they have no cell walls.

How did viruses become half living? Only a guess is possible. One possible explanation is that archea expelled unneeded genes instead of disabling them after a beneficial mutation increased the efficiency of metabolism. The surface of marine water contains incredible numbers of viruses. Marine microbiologists estimate that every milliliter of marine water contains 50 million virus crystals. They estimate that marine viruses alone number ten times the amount of all living organisms. Virus numbers are sustained because about 20 percent of all marine microbes die each day from viral predation. However they originated, viruses are predators of microbes and, alternatively, parasites of metazoan organisms.

EUKARIA

Eukaria are single-cell organisms with chromosomes enclosed in one or more compartments within the cell. In contrast, chromosomes in archea and bacteria cells float within a cell that has no interior compartments. The principal compartment in eukaria cells is the nucleus. Other compartments are called organelles. Organelles are the modified cells of encapsulated bacteria within eukaria cells. Most eukaria cells have many organelles that contain many sets of chromosomes. The first eukaria cell with one nucleus and no organelles was the first complex organism.

Eukaria evolved about 1.5 billion years ago when one species of bacteria encapsulated another species with which it had an obligatory symbiotic relationship. The encapsulated bacteria retained its DNA and became a nucleus within the cell. Many species of eukaria made multiple encapsulations of bacteria. The encapsulated bacteria retained their cell integrity and their DNA. Almost certainly, the evolution of bacteria into eukaria happened many times because symbiotic relationships are extremely common in contemporary microbes.

The largest eukaria cells in the contemporary biosphere have as many as 1,600 organelles and are 100 to 1,000 times larger than bacteria. Multiple encapsulations and multiple organelles containing DNA give eukaria a large pool of genes that can be continually transferred within cells. This transfer speeded the evolution of metazoan (multicell organisms) faster than the evolution of archea into bacteria during the first billion years of life (from four billion to three billion years ago).

How do mutations occur? Within eukaria cells there is a continuous transfer of genes between organelles and the DNA within the nucleus. Genes ensure that when cell divisions occur an exact copy of DNA is transferred to the new cell, however, the transfer of genes is never exact because genes are continually reshuffled on chromosomes. In humans, reshuffled genes are responsible for different colored hair, eyes, skin, height and other differences of greater significance, like brain capacity.

Gene shuffling also produces lethal mutations. For example, a lethal mutation in lions would be feet without claws and a beneficial mutation in marine iguanas in the Galapagos Islands is a flattened nose to aid grazing on algal mats growing on rocks in the tidal zone. If serial mutations increase the ability of an organism to gather food or reproduce, natural selection will eliminate less competitive members of a species and a new species will evolve.

Lethal and beneficial mutations of genes have identical origins. They are random occurrences that happen at random intervals and at random sites on DNA molecules. Natural selection is a biological sieve that rejects lethal mutations and preserves beneficial mutations, but it operates only after random mutations have occurred. In other words, mutations of genes and natural selection are unrelated processes. The speed and direction of evolution has no design, intelligent or otherwise. There is no divine plan. Evolution depends on random mutations that are beneficial.

Cell reproduction in eukaria is fundamentally different from bacteria. Bacteria reproduce by simple division (mitosis). Mutations are infrequent and beneficial mutations are therefore even less frequent. Eukaria cells reproduce by two divisions that produce four cells (meiosis). In the first division, chromosomes within the parent cell divide into pairs, and in the process genetic material is exchanged. Then the two successor cells divide into four cells. The four successor cells have half of the chromosomes of the parent cell, but in the exchange of chromosomes each new cell receives slightly different sequences of genes on DNA. The cumulative effect of these small, but continuous, differences in DNA sequences is more mutations; and if a serial sequence of mutations is beneficial, a new specie evolves.

For the first 3.3 billion years of life on earth, life existed as single-cell organisms (pre-archea, archea, bacteria, eukaria). It took about one billion years for pre-archea and archea to evolve into bacteria. An additional 1.5 billion years (to about 1.5 billion years ago) for some bacteria to evolve into eukaria. There was no certainty that diversity of life would proceed beyond single cell organisms, but it did.

It took an additional 800 million years for eukaria to evolve into metazoan organisms. The first metazoan organisms appeared about 700 million years ago. From another perspective, the mutations that began after the origin of life (about four billion years ago), to when archea evolved into bacteria (about three billion years ago), and bacteria evolved into eukaria (about 1.5 billion years ago), are far more complex than the mutations that have taken place during the last 550 million years when Cambrian chordate worms evolved into fish, reptiles and humans.

METAZOA

Metazoans are multicell organisms. They evolved from colonies of eukaria cells. All metazoan organisms use only eukaria cells in their bodies. A reservoir of genes existed in eukaria cells because most of those cells had encapsulated one or more archea, bacteria or other eukaria cells with which they had symbiotic relationships. The array of genes in the nuclei and organelles of eukaria cells was a reservoir of genetic possibilities for evolution into metazoa. It is highly probable that many species of colonial eukaria evolved into metazoan organisms.

Metazoan organisms evolved after DNA could sequence the activity of multitudes of enzymes that could synthesize other enzymes. These enzymes synthesized the specialized cells required for the evolution of eukaria into complex organisms. Thereafter, metazoa evolved into animals and plants. Metazoans that evolved into animals evolved the ability to control movement. Metazoans that evolved into plants encapsulated photosynthetic bacteria. Fungi were the third metazoan organism to evolve from undifferentiated metazoa. Like predatory animals, fungi metabolized an existing food supply that consisted of dead microbe cells that fell from the surface layer of photosynthesis to the ocean floor.

Metazoan life began when symbiotic eukaria cells failed to disperse after reproduction. They formed colonies that evolved interdependence. Interdependence was accomplished by an exchange of genes within eukaria cells. Interdependence then evolved into organisms that had an obligate colonial organization, meaning single cells from these colonial organisms could reproduce a whole organism. These metazoan organisms had to be simple because there were no cells that specialized in reproduction, metabolism or protection. In the contemporary biosphere, sponges retain the ability of one cell to reproduce the whole organism. The ability of one cell in these organisms to reproduce the whole organism is clearly a transitional stage to the evolution of organisms with greater complexity.

The time of the first appearance of metazoans is highly circumstantial. The most likely estimate is about 700 million years ago. The best evidence is the preservation of hydrocarbons in pre-Cambrian sediments (especially in Australia) that contain compounds (terpenoids) specific to eukaria. Microscopic metazoans were the likely source of these hydrocarbons. The best candidates for the earliest metazoan organisms are several species of very simple algae (volvocines) that have survived into the contemporary biosphere. They have only four to thirty-two cells glued together by gelatinous substances. They are living fossils from the pre-Vendian era.

Late in the Vendian era (about 560 million years ago) some soft-bodied metazoan organisms evolved into highly visible megaorganisms—this is known as the ediacaran assemblage. These organisms lived in shallow water on shelf sediments. They were anchored to bottom sediments by holdfasts in order to prevent dispersal by tides and to retain access to light. Many of them had fronds as long as one meter and some had multifronds. Other organisms were oblong disks that were as large as one meter in the long dimension. Others may have been shallow cups up to 25 centimeters in diameter that were flattened by fossilization. Ediacaran fossils have a worldwide distribution but none of them survived into the Cambrian.

The ediacaran assemblage is enigmatic because of the large size of organisms when all other late-Vendian organisms are barely visible with low-powered microscopes, as well as the inability to classify the assemblage as plant or animal. They were probably filter-feeding animals or were animals like some coral species that have an obligate symbiotic relationships with: (1) photosynthetic bacteria, (2) photosynthetic eukaria, (3) photosynthetic metazoans. They would be a kind of marine lichen.

Geologists, paleontologists and biologists have documented six events in the evolution of metazoan organisms during the Vendian era (610–550 million years ago): (1) most organisms were microscopic except for the enigmatic super giant fossils of the ediacaran assemblage; (2) colonial eukaria evolved into metazoan organisms after; (3) some colonial cells evolved specialized functions; (4) that became obligate for the organism's survival; (5) most Vendian metazoan organisms were microscopic and lived in niches where they evolved into animals, plants, fungi and other less familiar organisms; (6) fossils of these organisms are very scarce in late Vendian sediments, but there is very strong circumstantial evidence that the organisms existed. The evidence is the Cambrian explosion.

SEXUAL REPRODUCTION

Eukaria made sexual reproduction possible because all cells had two or more internal compartments that contained molecules of DNA. Eukaria cells continually exchanged genes between the nucleus and internal compartments (organelles). During meiosis, DNA is not exactly copied because DNA comes from two different cells that have variant genes. This is the basis for sexual reproduction (male–female). Sexual reproduction produces frequent mutations. All are random and most of them are lethal, but sexual reproduction has the possibility for increased numbers of beneficial mutations.

Mutations that produce specialized cells can increase the size of organisms, increase protection from predation, improve reproduction and improve metabolism. These mutations directly translate into the ability of metazoans to occupy and reproduce in vacant niches. During the first 150 million years of their existence (700–550 million years ago), metazoans were the largest marine organism. Even if they were composed of only four eukaria cells, they were giants when all other life in the oceans took the form of single-cell microbes.

After metazoans evolved there were few constraints on the shape of animals (body plans) that evolved to fill huge numbers of vacant niches. Circumstantial evidence strongly indicates that the hugely diverse body plans that became visible in the Cambrian evolutionary explosion evolved late in the Vendian era (555–550 million years ago). Measured against the previous record of bacterial evolution by mitosis, the evolution of eukaria and metazoa by meiosis and sexual reproduction proliferated new body plans. Although, there are no undoubted animal fossils in the Vendian era, almost certainly the highly variable body plans that became visible in the Cambrian evolutionary explosion were propelled by mutations that occurred because of sexual reproduction during the Vendian era.

SUMMARY

All of the evidence presented in this chapter indicates that evolution by natural selection is a hypothesis with a very high degree of certainty. There is no reason to believe that natural selection operates any differently now than it operated during the Precambrian era. Everything that science has discovered about the ability of bacteria and metazoa to reproduce and mutate is in agreement with Charles Darwin's hypothesis that natural selection propels evolution.

There is no evidence in the fossil record that there has been a pattern of beneficial mutations that guides evolution for any other purpose than immediate advantages in reproduction. The fossil record is a record of failed body plans, noncompetitive metabolism, insufficient protection, but above all, a failed ability to competitively reproduce. The fossil record clearly indicates that there is no linear pathway to reproductive success.

We know with a high degree of certainty that the molecular chemistry of RNA–DNA used by archea and bacteria for reproduction is the same pathway used for the reproduction of all cells in all metazoans, including humans. In the process of reproduction, life has evolved many forms, some of them bizarre, but there are some that have undergone minimal mutations for two billion years.

It took one billion years for pre-archea organisms to evolve into archea and archea to evolve into bacteria. It required an additional billion years for bacteria to evolve into eukaria. For the next 1.3 billion years (2 billion years to 700 million years ago) the molecular chemistry of eukaria cells continued to evolve. When metazoan organisms evolved about 700 million years ago, single-cell organisms had existed on earth for about 3.3 billion years. Metazoan organisms were late arrivals in the biosphere. In other words, 83 percent of the time that life has existed on earth it has been single-cell organisms.

More than 99.9 percent of all visible animal and plant species that have lived since the Cambrian era are extinct, but the survivors from the Cambrian eras have conserved a reservoir of genetic possibilities for replacing extinct species in subsequent geologic eras. The fossil record provides unambiguous evidence that extinctions are an integral part of the evolution of life and extinctions are ongoing events. The other side of the coin is that reproduction is the only purpose of life.

There is overwhelming scientific evidence that large, complex organisms like humans are evolutionary appendages to the world of microbes (archea, bacteria and eukaria). In the four billion years that life has existed, humans are one among many organisms that have temporarily occupied the top of the food chain. The post-Cambrian geologic record indicates that large animals, like humans, occupy the top of the food chain for a limited number of years and then become extinct.

This is contrasted to the Genesis myth that god created all organisms in the immediate past, in their exact contemporary forms, for the benefit of humankind. Or alternatively, god created life in the distant past and designed evolution so that humans would inevitably reach the apex of the food chain. In both versions, the Genesis myth is presented as incontestable certainty because the creation of life is the work of an omnipotent god. In both interpretations of the origin of life, humans are the ultimate purpose of life on earth. There is zero scientific evidence for the Genesis account of the origin of life or for the purpose of human life on earth.

This interpretation, however, conforms to a strong human bias to believe that the contemporary status of humankind at the top of the food chain was designed by god. Humans could then glorify him for the gift of life and give thanks for enjoying supremacy over the earth's biosphere. This interpretation of the origin and purpose of life is wholly based on deductive reasoning that claims that the Genesis myth is a true explanation for the origin of life.

CHAPTER 2

DIVERSITY OF LIFE

This chapter documents the evolution of metazoan life and its reproductive ability from the beginning of the Vendian era, 610 million years ago, to the present. Diversity of life is dependent on occupying the maximum number of niches where life can reproduce. Diversity of life is a consequence of evolution. Prior to humans becoming super predators (during the past 40,000 years), the earth probably had as much biodiversity as any time in the geologic record of life. Whenever there are long periods with a benign biosphere, plants and animals rapidly evolve to fill the available niches. The geologic record has had many periods, often prolonged, when severe conditions have impacted the biosphere. During these periods of stress, the number of species diminishes and the purpose of life comes into very clear focus: ability to reproduce.

VENDIAN

The first good evidence for the appearance of metazoan organisms is about 700 million years ago in the pre-Vendian era; although, metazoa could have evolved 200 million years earlier (900 million years ago). From their estimated origin 700 million years ago, metazoan organisms have existed for only 17 percent of the estimated time that life has existed on earth. When metazoan organisms appeared, they were newcomers in the evolution of life.

The fossil record of the Vendian era has three very different categories of organisms that are difficult to correlate, but they share one attribute—they lived in the photic zone of shallow marine water. The first category is photosynthetic bacteria that evolved from archea. They provided food for most metazoan marine animals and fungi in the Cambrian and continue to do it in the present. The second category is marine animals. Most were microscopic because few had evolved visible shells or exoskeletons; but compared to photosynthetic bacteria and other single-cell microbes, they were giants at the top of a short food chain. Although nearly invisible, they were hugely abundant.

The third category is the ediacaran assemblage. As noted in the previous chapter, it evolved near the end of the Vendian era (about 555 million years ago). It was composed of super large organisms and they occurred worldwide. They are enigmatic in the evolutionary tree of life because they cannot be clearly identified as plants or animals.

Most had a flat morphology. Some were quilted disks (flat worms) that apparently moved on bottom sediments. They could have been scavenging animals. Others were upright fronds as long as one meter, with a holdfast to bottom sediments. Fronds maximized the acquisition of sunlight in shallow water. They also maximized opportunities for filter feeding in tidal currents. Another plausible interpretation is that the fronds were animals that hosted symbiotic photosynthetic algae in return for a platform for reproduction.

Near the end of the Vendian and into the earliest Cambrian eras (550–545 million years ago), a few species of microscopic arthropods, mollusks and sponges began to biomineralize spicules, scales and blades (sclerites) to embed in their outer membranes. Biomineralization evolved to increase protection from predation. The spicules, scales and blades that became part of the fossil record were 0.02 millimeters, a fraction of the thickness of a human hair, and require a low-powered microscope to see, but there is no doubt about their purpose. Other sclerites may have been mouth parts (rasps) of predatory worms that were used to pierce the thin, fragile shells and exoskeletons of arthropods, brachiopods and other animals that evolved in the late Vendian and became visible in the Cambrian evolutionary explosion.

Biologists assign the microfossils of the late-Vendian era to the animal kingdom because a high percentage of visible organisms that were fossilized in Cambrian sediments had varying degrees of mobility. Put another way, during that time frame the invisible and barely visible animals of the Vendian era were competing to be first occupiers of vacant niches for metazoan animals, and mobility was a huge survival advantage.

In an ocean where there were no large mobile predators, size protected the ediacaran assemblage from excessive predation. This radically changed at the Vendian–Cambrian boundary. Size was not an effective defense against swarms of mobile worms longer than one millimeter that grazed on bacteria/algae in the fronds and on algal mats, just as size was not an effective defense for stromatolites anchored to the ocean floor. The ediacaran assemblage of super giant organisms was extinct by the beginning of the Cambrian era (550 million years ago); and stromatolites survive into the present only in restricted environments where there is minimal predation from animals that graze on algae.

CAMBRIAN

At the beginning of the Cambrian era the benthic and pelagic zones were nearly empty of animals with visible shells and exoskeletons. Very large numbers of first occupancy niches were available. Fossil evidence indicates animals that survived the extinction of the ediacaran assemblage rapidly evolved into larger animals that filled benthic and pelagic niches. The abrupt appearance of visible fossil shells and exoskeletons is called the

Cambrian evolutionary explosion. Size and visibility occurred because there were few constraints on the evolution of metazoan animals. Many first occupancy body plans were bizarre, but they survived because any body plan was acceptable to fill vacant niches for larger animals.

The Cambrian explosion has three parts: it was an animal event, a mobility event, and a body size event. Most body plans conformed to three symmetries: bilateral (worms, arthropods), radial (jellyfish, starfish) and amorphous (sponges).

Other characteristics of the Cambrian evolutionary explosion are: (1) a huge increase in the number of visible animals; (2) that filled vacant niches; (3) biomineralization of shells and exoskeletons for protection from predation, which equally served for muscle attachment needed to increase mobility; (4) a high percentage of new animals had bilateral symmetry to facilitate movement, (5) bilateral symmetry was also essential for animals to increase their size; (6) most plants continued to be one-cell marine organisms.

There are two sources of evidence for the Cambrian evolutionary explosion—visible fossils and new species. The sudden appearance of large numbers of visible fossils compared to the Vendian era is documented in the Burgess shale (Canada), Chengjiang shale (China), Sirius Passet shale (northernmost Greenland), and sediments in Utah in the United States. Fossils from these localities consist of shells, exoskeletons, spicules and carbonaceous films of animals without shells. Shells, spicules and sclerites were made of calcite (calcium carbonate), apatite (calcium phosphate) and opaline silica (silicon dioxide). Shells probably evolved from fused sclerites embedded in the outer membranes of late Vendian animals. In addition, chitin (a tough organic armor, such as that which encases crabs) was used to make exoskeletons.

By the mid-Cambrian era many animals were fully encased

in shells and exoskeletons; and these were frequently covered with spines. In addition, small amounts of magnetite, a hard iron oxide, were used for specific purposes by both bacteria and metazoan organisms.

Although shells vastly increased chances for fossilization, most Cambrian animals continued to have soft bodies. There are several localities in the world where bedding planes in fine-grained shale preserve carbonaceous films of soft bodied Cambrian animals. These films often preserve internal anatomies in exquisite detail. These details are often sufficient to allow Cambrian animals to be compared with animals in subsequent geologic eras, and into the present. Again, the best preserved carbonaceous film fossils are in the Burgess shale in Canada, Greenland's Sirius Passet location and China's Chengjiang shale.

Shells and exoskeletons in the lowermost Cambrian sediments in China are very small and thin, but there is a very large diversity of forms (species). Shells and exoskeletons from Chinese sediments are recovered by soaking carbonate nodules recovered from black shale in a bath of very dilute acid (usually 5 percent acetic acid). These fossils vary from 0.7 to 6.0 millimeters long. Some are cones, others are tubes, and all of them are very fragile.

During the rest of the Cambrian era the size of animals with shells and exoskeletons increased in size. Biomineralization, mobility and size proceeded together in the Cambrian explosion. By the middle Cambrian, fossils recovered from the Burgess and Sirius Passet shales are up to two centimeters long, and the largest predator was 60 centimeters long.

Did Cambrian animals evolve size because vacant niches were available? Probably. Did Cambrian animals evolve size because some microanimals had lucky mutations? Probably.

Almost certainly, the mobile predators that evolved at the Vendian–Cambrian boundary exterminated a high percentage of

Vendian organisms that lacked mobility, lacked the protection of shells or spines, or had body plans that were noncompetitive for feeding and reproduction at a time when predatory animals were rapidly evolving in size and mobility.

In the early Cambrian ocean any body plan was acceptable to fill vacant niches for large animals. Paleontologists estimate that as many as 100 different body plans evolved in the Cambrian era. Many of the body plans were strange. These body plans were not immediately lethal because vacant niches in the early Cambrian accepted any animal that could grow in size and reproduce regardless of body plans. These animals survived for varying lengths of time because they were first occupiers of vacant niches and faced minimal competition.

By mid-Cambrian era, animals with first occupancy body plans were experiencing intense competition. Natural selection began to cull animals with the least efficient body plans for feeding and reproduction. Fossils from the Cambrian evolutionary explosion are the necessary beginning for understanding the origin of diversified life in the contemporary biosphere.

The most successful animals that evolved in the Cambrian era had bilateral symmetry. Bilateral symmetry aided movement of animals, especially those with a worm morphology, because it increased the efficiency of swimming. Swimming by lateral or vertical body movements (round worms, flat worms) helped escape predation and/or move to food sources (algal mats, concentrations of organic detritus, small worms to be eaten).

Bilateral symmetry coupled with forward mobility concentrated ingestion of food at the front end of an animal—in the direction it was moving. It also concentrated metabolism in one stomach. Food entered at one end (mouth), and waste was expelled at the other end (anus). Concentrating metabolism in one stomach facilitated metabolism so that energy could be distributed to muscles that needed high energy inputs for movement.

Neural systems evolved to control the movement of swimming muscles.

During the Cambrian era, animals with first occupancy body plans had variable numbers of beneficial mutations. Those with insufficient beneficial mutations were squeezed into smaller and smaller niches. Rapidly changing sizes of niches and evolution of larger animals that specialized in predating specific species of smaller animals reduced reproduction opportunities for these animals. There was no way to predict what mutations in feeding, mobility and size would preserve a species from extinction.

Nevertheless, Cambrian animals with diverse body plans established a food chain that was similar to the contemporary marine food chain. The foundation of the food chain in the Cambrian ocean was the same as in the contemporary biosphere but much shorter. The foundation of the food chain was photosynthetic bacteria and the apex was predatory animals.

In the remainder of this section, I will focus on Cambrian animals with bilateral symmetry because, from the end of the Cambrian era (495 million years ago) to the present, animals with bilateral symmetry have evolved into highly diverse marine and terrestrial species. Animals with amorphous symmetry (sponges), those with radial symmetry (starfish, jellyfish, coral), have remained small compared to animals with bilateral symmetry. I ignore these animals. I will concentrate on chordates because they evolved fish, reptiles and mammals; and also arthropods because they evolved into the largest number of species of land animals (insects).

WORMS

Worms were newly visible animals. They are preserved as carbonaceous films that frequently preserve internal organs. They had bilateral symmetry and neural systems that controlled muscles to move them to food sources and to escape predation. Neural systems were essential to control muscles in order to actively seek food as predators or escape predation, or graze on algal mats, or scavenge organic detritus in bottom mud. The largest worms actively hunted small metazoans, including the smallest worms. The increased size of animals changed the structure of shallow water sediments because worms burrowed into them to a depth of several centimeters in order to escape predation. Bottom sediments no longer had thin, regular bedding. They were biodisturbed.

Worm morphology was a first occupancy body plan in the Cambrian evolutionary explosion. At the beginning of the Cambrian era most animals with worm morphology were less than one centimeter in length. Animals with worm morphology had two body plans: those with notochords and those without. Notochords are elastic rods of muscle along the length of a worm's body. Muscles attach to them so that worms with notochords became efficient swimmers. Swimming efficiency was strongly aided by a tail and a body shaped like a vertical or horizontal ribbon.

Chordate animals have a nerve cord that parallels the notochord in order to control swimming muscles. Both the notochord and nerve chord are along the back. Chordate animals preserved as Cambrian film fossils had a worm morphology, but they were not worms. Chordate animals belonged to a separate family that happened to have a worm morphology in the Cambrian. The body plan of chordate worms maximized mobility, and this translated into more opportunities to gather food and use speed to escape predation by larger animals. Most chordate worms, however, hugged the ocean floor or dived into bottoms muds, and competed with obligate seafloor dwelling animals for food.

The notochord of chordate animals was the basis for two

mutations. It became the template for the biomineralization of vertebrae that formed a spine that encased the dorsal nerve cord. Vertebrae bones evolved bone extensions that became ribs for attaching strong swimming muscles.

The combination of sinuous body motion, tail propulsion and control of body movements increased the speed of movement for chordate worms. They filled a niche in the pelagic zone (ocean surface) far above benthic animals that inhabited the ocean floor. From the vantage of height they could see and capture benthic animals. Several chordate worms evolved a mouth appendage with a rasp termination that could bore holes in the shells of benthic animals. They could also feed on plankton in the pelagic zone.

The mobility of chordate worms was a huge evolutionary success. Beginning about the middle Cambrian, chordates with a worm morphology evolved in a radically different direction than true worms (annelid worms) that have nerve cords along the undersides of their bodies. The evolution of biomineralized notochords into bone with rib extensions are the ancestors of all marine and terrestrial vertebrates, including humans.

ARTHROPODS

About 80 percent of all living animals described by biologists are species of arthropods. Today, arthropods are the most numerous visible animals. Most of them are insects and the most common are ants. Others are spiders, scorpions, crabs and lobsters. Arthropods were also the most common animals preserved as carbonaceous film fossils and exoskeletons in Cambrian sediments. They are about 35 percent of the species preserved in the Burgess shale.

By the middle Cambrian many arthropods had increased

their size to several centimeters. Most were scavengers that fed on the rain of dead bacteria that fell from the pelagic zone. Many benthic scavengers evolved numerous body segments with each segment having pairs of jointed appendages (legs). Legs gave them mobility to move on soft sediments that composed the ocean floor. The appendages often had feathered gills attached to them that extracted dissolved oxygen from water. A few, however, were swimmers that actively hunted food in the pelagic zone.

Besides worms and arthropods, the surviving body plans from the Cambrian were mollusks, sponges, jelly fish and annelid worms. The body plans of mollusks, sponges, and jelly fish were efficient for filtering pelagic bacteria and other microorganisms from marine water and conveying them into a mouth and stomach.

Animals that survived competition in the Cambrian evolved shells, swimming mobility, leg mobility, eyes, poisons and camouflage to escape predation. By the end of the Cambrian, substantial numbers of species of arthropods had chitin exoskeletons, some with spikes, to protect their bodies from predators attacking from above. Undersides were not protected in order to allow jointed appendages to move on bottom sediments and aid in feeding.

Natural selection does not produce perfection. The fossil record tells us that most Cambrian animals were only temporary successes because 90 percent of the visible animals with first occupancy body plans were extinct by the end of the Cambrian era. The direction of natural selection is to preserve beneficial mutations that allow organisms to increase efficiency of reproduction and metabolism in order to occupy vacant niches that are immediately available. Extinctions tell us that beneficial mutations confer only temporary advantages because the dimension of these niches is in continual flux.

The ability to reproduce and metabolize new sources of food is always conditional because many organisms are simultaneously undergoing mutations. In a biosphere governed by natural selection, the timing of beneficial mutations increases the rate of survival for organisms that can immediately compete for available niches. Animal occupancy of niches is not permanent because beneficial mutations in other species continually modify the ability of animals to retain a place in their niche. For example, the largest Cambrian animal was a predator. It reached a length of about 60 centimeters (*Anomalocaris*). It was the giant among Cambrian animals. It was a swimmer that captured benthic animals from above but it was extinct by the end of the Cambrian.

The fossil record is overwhelming evidence that natural selection produces flawed organisms because the fossil record is a record of extinctions. Most species have an indefinite time of existence before extinction; and there is a high degree of certainty that most species will become extinct. This has not changed since life's inception.

The mass extinction at the end of the Cambrian era was probably due to intense competition among animals with first occupancy body plans. These animals were adequately efficient to feed and reproduce in vacant niches. They became extinct when they were unable to compete with other first occupancy animals that evolved more efficient body plans, more efficient means of protection, more efficient metabolism, mobility and reproduction. Some body plans, however, were evolutionary successes. The five most successful (and visible) were worms, arthropods, brachiopods, mollusks and sponges.

Virtually no new body plans have evolved since the mass extinction at the end of the Cambrian era. The body plans of animals that survived have been strongly conserved to the present, but have undergone intense elaboration in subsequent geologic eras. During the 200 million years between the early Cambrian (550 million years ago) and 350 million years ago, the genes that evolved exoskeletons in arthropods and notochords in Cambrian worms evolved into flying insects, fish with scales, animals with jaws and teeth, and marine and land reptiles that had internal skeletons of bone.

ORDOVICIAN

The first fish evolved at the end of the Cambrian era or near the beginning of the Ordovician era (495 million years ago). They evolved from chordate worms. The first fish had armored carapaces but no skeletons or jaws. Swimming muscles attached to the carapace and thick scales covered the rest of their bodies. The earliest fish fossils were about 10 centimeters long but they rapidly increased in size. Most were probably bottom feeders by sucking in small worms and plankton. They had small eyes and their heavy carapace and heavy scales limited movement in open water. During the Ordovician at least fifteen body plans of fish evolved.

Fish did not dominate the benthic zone during the Ordovician. The benthic zone was dominated by sea scorpions (Eurypterids) that grew as long as two meters. They are the largest arthropods in the fossil record. By the end of the Ordovician they were extinct. Almost certainly, Eurypterids failed to survive because they failed to compete for food with highly mobile fish. Fish achieved superior mobility by evolving out of their armored carapaces, in a similar way that squid evolved out of their shells, and sharks evolved out of their bone skeletons. The superior speed and mobility of fish were highly advantageous in acquiring food and escaping predation.

The Ordovician ended 438 million years ago with a mass

extinction of unknown cause. The most common explanation is large-scale continental glaciation coupled with a substantial lowering of ocean levels. The geological record indicates that glaciation was widespread during the Ordovician because major continental land masses drifted over the South Pole. A meteor impact is a possibility, but there has been no concerted effort to look for sediments that span the Ordovician–Silurian boundary in hopes of finding stressed quartz.

More likely, the mass extinction occurred because of competitive displacement by many animals that were evolving new organs that allowed them to fill niches that were inadequately filled. Among the beneficial mutations were fish with jaws. The earliest fossils of jawed fish are found in the early Silurian (435 million years) and by late Devonian (365 million years ago) they replaced most jawless fish. At the same time, the largest fish had risen to the top of the food chain. Fish with skeletons of bone were swift swimmers that were capable of darting into bottom mud to capture animals and then escape into the benthic zone. Squid achieved the same speed and mobility by propelling themselves with jets of water.

By the Devonian era (400 million years ago) fish evolved to one to two meters long and were the largest marine predators. By the end of the Devonian (362 million years ago), there were over 200 genera of fish with the largest being six meters in length. They were at the top of the marine food chain.

Implausibly, Cambrian chordate worms that evolved bone skeletons became top predators in several food chains in subsequent geologic eras. They did this by evolving jaws with teeth, breathing air, and fins into legs and wings. Some of these animals were sharks, dinosaurs, birds (raptors), lions and humans.

LAND NICHES

Land colonization by metazoan plants and animals began about 445 million years ago, near the end of the Ordovician era. Microbial life was abundant, but there was no visible life. The most favorable habitat for the evolution of land plants and animals was tidal zones. From there, both plants and animals moved inland as they evolved ways to survive desiccation. Metazoan plants first colonized fresh water in inland lakes and swamps. There they were a source of food waiting to be exploited by animals that could reproduce in fresh water. In order to access this food, animals had to evolve ways to survive desiccation, breathe oxygen and reproduce on land.

In the 76 million years from the beginning of the Silurian era (438 million years ago) to the end of the Devonian era (362 million years ago) plants underwent an evolutionary explosion similar to the evolutionary explosion of animals during the Cambrian era. The first visible land plants were mosses because they evolved a way to reduce water loss when permanently exposed to air. They appeared in the early Silurian (435 million years ago). The earliest fossilized mosses were two to three centimeters high with single stems or single stems with branches. They formed mats along riverbanks and in moist habitats that were frequently flooded by rain.

During the Devonian era (417–362 million years ago) plants rapidly evolved to occupy a high percentage of land niches. Successful species were those that maximized the acquisition of sunlight and could retain water within plant tissue during periods of desiccation. Competition for light put a premium on a canopy of leaves that kept competitors in the shade. Growing upward did this, but upward growth required strong central stems (trunks).

Tall trees had to survive high winds and carry water to leaves where photosynthesis took place. Tree trunks evolved into bundles of stiff, hollow, flexible fibers that swayed in normal winds but did not snap. By middle Devonian (385 million years ago), tree ferns formed a dense cover in favorable habitats and several species grew eight meters high. By late Devonian (375 million years ago) trees reached twenty meters. At the end of the Devonian the land surface was covered with very large numbers of visible plant species.

It took 3.5 billion years from life's inception (to the beginning of the Cambrian era) to evolve visible marine animals but only 76 million additional years to evolve visible terrestrial plants and arthropods to feed on them; and complementary arthropods (spiders) to feed on the arthropods that ate plants. The evolution of microscopic plants into visible terrestrial plants is an event that occurred in only 2 percent of time in the geologic record of life.

The path of evolution of the first visible plants into large plants is better understood than the evolution of microscopic marine animals into large marine animals, and the further evolution of marine animals into land animals. Plant evolution is better understood because very large areas of post-Silurian continental sediments contain huge amounts of plant fossils. These fossils have been preserved as coal.

At the same time plants were evolving to fill huge numbers of land niches, some marine arthropods were evolving into fully terrestrial arthropods (mostly insects) to feed on fully terrestrial plants. These arthropods were the first animals to colonize land because, like plants, they evolved ways to prevent dehydration and, at the same time, breathe air through microtubes in their exoskeletons. Arthropods are the first undoubted terrestrial animals. Their fossils have been recovered from sediments that are about 400 million years old. About two million years after arthropods evolved the ability to breathe air, some species evolved into insects with wings in order to reach new sources of food.

One of the first plant defenses against animal predation was spines on their stems. These spines later evolved into leaves. Later defenses were poisons, sticky sap and symbiotic relationships with insects that repelled insect predators.

AMPHIBIANS

Amphibians evolved from fish that fed in shallow marine water. Many fish in the contemporary biosphere occasionally gulp air when water has a low content of dissolved oxygen. In the late Devonian (365 million years ago), several species of these fish evolved organs that allowed them to move onto land. They evolved lungs to respire oxygen directly from the atmosphere and legs for locomotion. These organs gave them access to a very large food supply in many unoccupied niches. Contemporary lungfish living in Africa, Australia and South America are living fossils of this transition.

All contemporary amphibians are obligate land animals but continue to use water for reproduction, but it is fresh water. Some amphibians continue to reproduce like fish. Females lay eggs in lakes, rivers, ponds or pools and are fertilized by males ejecting a cloud of sperm. Other amphibians, like frogs, have evolved a larva stage in the form of a fish with gills (tadpoles). Other amphibians (salamanders) have live births but the newly born live in water and have long, feathery gills for respiration until they develop lungs. When lungs are fully functional they move onto land in search of food.

For about 10 million years in the lower Mississippian era (360–350 million years ago), amphibians were at the top of the terrestrial food chain. Some species grew three meters long.

It is probable that only one family of amphibians survived the mass extinction at the end of the Permian (251 million years ago) because only one family has survived to the present. It is the family of frogs and salamanders, both of which are predators. Amphibians that survived the Permian mass extinction were small and survived by retaining reproduction in fresh water. Their principal survival strategy was to conceal themselves in burrows, under rocks in flowing streams, in aquatic foliage or in damp forest litter.

Contemporary salamander species that live in forest litter have evolved some of the anatomy of reptiles. Male sperm is in a packet that females pick up and store in a pocket. Fertilization of eggs is internal and eggs are large and laid at appropriate times and in optimal locations. When eggs hatch they are miniature adults. Other contemporary species of salamanders and frogs have lost their lungs. Respiration is totally through the skin. Skin respiration is possible because they are small, the skin is always wet and permeable, and they have body plans with large surface areas.

REPTILES

Reptiles evolved from amphibians. The first reptilelike animals entered the fossil record during the early Mississippian era (360 million years ago). They filled a huge number of land niches because amphibians could not reproduce away from bodies of fresh water.

The key mutation that evolved reptiles from amphibians was eggs with waterproof shells. Reproduction became possible in highlands distant from permanent sources of water. Waterproof shells made it possible for reptiles to have a total terrestrial existence. This greatly increased the number of feeding and reproductive niches.

Impermeable shells required that fertilization take place before the egg was enclosed, and the energy required to produce an enclosed egg meant that fewer eggs were produced and they were larger than amphibian eggs. Reproduction with shells, however, did not totally escape the marine ancestry of reptiles because the water that bathes embryos within shells has the chemistry of seawater.

The success of herbivorous reptiles was dependent on developing a symbiotic relationship with bacteria that could metabolize cellulose. Cellulose is the principal constituent of woody plants; and it is extremely resistant to biodegradation. There are, however, bacteria that produce an enzyme that does it quickly. These are the same bacteria that live in the guts of termites. Herbivorous reptiles ingested these bacteria and evolved a pouch for them in their gut. In return for protection in the guts of reptiles, these bacteria metabolized the cellulose that browsing reptiles brought to them in large quantities. The waste products of the biodegradation of cellulose are sugars and starches that herbivorous reptiles metabolized for food.

This symbiosis allows these bacteria to survive and multiply without predation. It also allowed some herbivorous reptiles to grow to great size because they had a continuous supply of high energy food. Herbivorous reptiles that became part of the Permian fossil record ranged in length from 0.4 meters (rabbit-size) to 3.5 meters (rhinoceros-size). At the end of the Permian era (251 million years ago) a modern terrestrial food chain was functioning but it was severely stressed by the mass extinction that ended the Permian era.

More than 80 percent of all families of terrestrial vertebrates became extinct in the Permian mass extinction. The largest reptiles that survived the Permian mass extinction had bodies about the size of a large dog. Several families of mammal-like animals also survived but they were outnumbered by more families of reptiles. The surviving families of reptiles had larger bodies and occupied many more niches than mammal-like animals; and during the next three geologic eras (Mesozoic), reptiles became the dominant land animals.

In the lower Cambrian, new species with diverse body plans appeared with explosive suddenness. This did not happen after the Permian extinction in spite of the very large numbers of vacant niches. No new body plans evolved to fill vacant niches because a sufficient number of species of marine and terrestrial animals survived. Surviving species flooded vacant niches. This gradually changed during the lower Triassic (251 to 245 million years ago) because new species evolved to fill smaller and smaller niches.

By the late Triassic (210 million years ago), diverse species of herbivorous and carnivorous reptiles had evolved to fill most marine and terrestrial niches. Diversity returned to the biosphere. Put in different words, species that evolved in the Triassic era and in subsequent geologic eras were elaborations of body plans that had survived the Permian mass extinction.

Fossil evidence indicates that during the next 186 million years, between the Permian mass extinction and the mass extinction at the end of the Cretaceous era, there was an evolutionary race between herbivores and carnivores. After the mid-Triassic (230 million years ago) many herbivores and carnivores rapidly evolved larger bodies. Large herbivores evolved armored skin with embedded bone plates, long whip tails (sometimes with bone spikes or spiked maces on the ends), social behavior to protect nesting sites and herding behavior for protection while browsing.

Carnivores also evolved large size. Many also evolved bipedalism for lunges and large heads full of serrated teeth for tearing flesh. Smaller carnivores also evolved bipedalism to gain speed for running down small herbivorous reptiles or to escape larger bipedal carnivores. They also adopted social behaviors by hunting in packs and for protecting nesting sites. In the 186 million years from the beginning of the Triassic to the Cretaceous mass extinction (65 million years ago), three counterbalancing mutations governed the evolution of land animals: (1) increasing size for both herbivores and carnivores, (2) bipedalism for carnivores and whip tails for herbivores, (3) herding for herbivores.

Among the largest herbivores was *Apatosaurus* (*Brontosaurus*). It had a very long whip tail that gave it a length of twenty-seven meters and an estimated weight of twenty-eight tons. Another herbivorous dinosaur, similar in body plan to *Apatosaurus*, weighed eighteen tons. Both of these dinosaurs were dwarfed by *Brachiosaurus* that had a body length of about twenty-five meters, including its tail, and it weighed an estimated fifty-five tons. These are the largest terrestrial vertebrates to have evolved on earth. The principal predator that fed on these giant herbivores was *Tyrannosaurus rex* and other species in that family. *T. rex* was a bipedal carnivorous dinosaur that was about fourteen meters long and weighed 7.5 tons when mature.

MAMMALS

Mammal-like reptiles evolved from reptiles soon after reptiles had evolved from amphibians. The evolution of reptiles into mammals is well documented because many intermediate families have been preserved in the fossil record. Mammal fossils are distinct from reptile fossils because their lower jaws are a single bone and their jaws have very different dentition. The first fossils of mammal-like reptiles are found in late Permian sediments (255 million years ago). Some mammal-like animals survived the Permian mass extinction (251 million years ago), but not until the late Triassic (210 million years ago) were there unmistakable mammals. After the Permian mass extinction, mammal-like animals appeared to have a competitive advantage over reptiles to fill vacant terrestrial and marine niches, but this did not happen. All the niches for mega-animals in the Triassic, Jurassic and Cretaceous eras (Mesozoic) were filled by reptiles.

What are the competitive advantages of contemporary mammals compared to contemporary reptiles? The four most obvious are warm blood (endothermic), body hair, blubber belts and feathers. Hair, blubber, and feathers are very efficient insulators for conserving body heat. A high, constant body temperature maintained by warm blood keeps mammal brains fully functional because they have first call on oxygen and energy transported by blood. This ensures an ability to respond to severe temperature fluctuations on land and an increased ability to escape predation.

Four less visible advantages are: (1) reproduction by live births, (2) suckling young with mammary glands, (3) periods of dependency after birth for young animals to learn survival skills, (4) larger and more complex brains in relation to body weight. These advantages confer a higher rate of postbirth survival on mammals compared to reptiles.

How were these advantages nullified by reptiles during the Mesozoic? There is no conclusive answer. For the most part, the mammal-like animals that survived the Permian mass extinction were small. Most were insectivores and other solitary carnivores. The largest were about the size of a beaver.

The best explanation is that, after the Permian mass extinction, most surviving reptiles: (1) were larger than most surviving mammal-like reptiles, (2) occupied many more terrestrial niches, (3) could evolve increased size in these niches, (4) rapidly evolved social behavior. These advantages neutralized most of the advantages of small mammal-like reptiles with solitary habits and were sufficient for reptiles to dominate almost all of the terrestrial and marine niches in the Triassic, Jurassic and Cretaceous eras.

Mammals, however, gradually increased their range in the Jurassic and Cretaceous eras because their high constant body temperature allowed them a high level of activity at cool temperatures. This translated into hunting and feeding at night when lower nighttime temperatures made reptiles sluggish. They could also survive and reproduce in cool climates because body hair conserved heat better than the scales, armor plates and bare skin of reptiles. Before the Cretaceous mass extinction, some low-weight reptiles evolved feathers from scales in order to conserve body heat. Feathers are very efficient insulation.

Nighttime sluggishness had less affect on larger reptiles because their bulk, in relation to the area of their skin, better retained body heat. Mammals in the Mesozoic era did not compete with large reptiles.

Herds of large reptiles are known to have migrated into higher latitudes during seasons of long sunshine (summer) but they could not remain there during cooler seasons (winter). Mammals could live the entire year in cool climates without reptile predation. In cool climates food resources during cold seasons were exclusively theirs. In warmer climates, however, the advantage of size conferred sufficient advantages for reptiles of all sizes to dominate terrestrial and marine habitats until the Cretaceous mass extinction 65 million years ago.

Among the mammals that survived the Cretaceous mass extinction were the first primates (lemurs, tarsiers, lorises) that lived in trees. Like all of the mammals that survived into the Cenozoic era, they were small and most were probably insectivores, like most lemurs today. Most species of these early primates became extinct on continental land masses because they were poor competitors with other primates (monkeys) that competed in their niche in the forest canopy.

Large number of species, however, survived until recently on the island of Madagascar because Madagascar rifted from the African continent early in the Cenozoic era. Humans did not colonize Madagascar until about 1,500 years ago. They rapidly destroyed favorable habitats, and many species of lemurs, tarsiers and lorises have become extinct in human memory.

Marsupials were the largest mammal family to survive the Cretaceous mass extinction. In Asia and Africa marsupials were gradually replaced by placental mammals. Replacement did not occur in South America and Australia where marsupials continued to be dominant mammals because both continents were giant islands. Placental mammals were poorly represented in South America and absent in Australia. In Australia, marsupials had no competition for food resources. Separation from other continental landmasses preserved marsupial dominance in Australia until humans arrived 40,000 years ago.

Placental mammals had reproductive advantages over marsupials because survival skills could be taught to fully formed newborns. By mid-Eocene (50 million years ago), placental mammals occupied almost all terrestrial niches in Africa, Europe and North America because these continents were joined.

The first flowering plants preserved in the fossil record are mid-Cretaceous (110 million years). Angiosperms are flowering plants and they reproduce by seeds. Foremost among the new families of flowering plants were grasses. Flowering plants require pollination in order to produce seeds. Grasses and some trees are pollinated by wind; but insect pollination is more efficient for most broad leaf plants because they are highly dispersed compared to grasses and trees.

In most habitats, seeds are a more efficient way of reproduction if they can be dispersed. In the post-Cretaceous recovery, the evolution of flowering plants was propelled by symbiotic relationships with nectar-eating insects that pollinated seeds, seed-eating animals that dispersed seeds and herbivorous animals that ate grasses. Feeding on grasses and other angiosperms prevented grasslands from becoming forestland.

CONVERGENCE

After the Cretaceous mass extinction, species of herbivores rapidly filled the niches vacated by reptiles because abundant food was available on grasslands and savannas. These habitats, however, offered little protection from carnivores. The herbivorous mammals that evolved to graze grasslands and browse savannas evolved the same five behaviors that herbivorous reptiles had evolved to escape excessive predation. They were body size, speed, armor, camouflage and herds. Size and herding were the most effective defenses against excessive predation. Almost all large mammalian herbivores live in herds because escape by speed is not an option. Herding offers protection for all but the old and weak.

Not only did mammals rapidly evolve to fill herbivore niches once occupied by reptiles but, like reptiles, they increased in size. By 35–33 million years ago the largest land mammal known in the geologic record had evolved. It was a species of rhinoceros that was five and a half meters tall, seven meters long, and it weighed between fifteen and twenty tons. It was more than twice as heavy as the largest African elephant. All species were extinct by five million years ago.

Rhinoceroses typify the evolution toward size of several families of herbivores. Giant mammals require much more food and this makes them vulnerable to climate changes, particularly desertification because this reduces food supplies. Rock paintings in the Sahara desert document that elephants, giraffes and rhinoceroses grazed on grasslands and savanna vegetation

10,000 years ago and humans hunted them. After European continental glaciers melted, there was a dramatic climate change. The grasslands and savannas of northern Africa were replaced by the Sahara desert because the postglacial climate in Europe changed the pattern of moisture laden winds. By 7,000 years ago herds of large mammals had disappeared from land that became the Sahara desert.

The fossil record clearly indicates that convergence of body plans was a way for unrelated families of animals to fill new niches. Many unrelated families of mammals evolved long legs with hooves that increased the ability to run on grasslands and savannas. Especially antelope evolved hooves and speed to reduce predation. Some species of antelope and deer also adopted herding in order to reduce excessive predation and maintain competitive reproduction. Other animals evolved legs that were posts to support the heavy bodies of megamammals (elephants) Other terrestrial mammals evolved legs into flippers (seals) and filled the vacated niches of marine reptiles.

Eyes are probably the best example of evolutionary convergence. Very small arthropods in the lowermost Cambrian had eye spots to detect movement of predatory worms or predatory arthropods. Eyes were not essential for feeding because the legs of Cambrian arthropods were covered with bristles and these bristles were covered with bristles. Arthropods fed by touch. They swept bacterial detritus from the surface of marine sediments into their mouths. Other animals, however, evolved eyespots and used them to locate food.

Humans are sight dependent. Smell, hearing and touch are significantly less efficient in humans than in most other mammals; therefore, sight has a particular fascination in understanding the evolutionary complexity that mutations can accomplish. The evolution of eyes is one of the best ways for readers to understand that evolutionary complexity is, ultimately, a product of natural selection. Early Cambrian animals evolved light sensitive cells (eyespots) for two purposes—escaping predations and feeding. Cambrian trilobites (arthropods) evolved lens of calcite (CaCO3) to detect movement of predators, and other Cambrian animals evolved cornea lens (organic) to aid feeding.

After the Cambrian era, more than forty different families of invertebrate animals evolved eye morphologies with different nerve circuitry. Two principle types of eyes evolved: (1) compound eyes, especially in insects, (2) enlarged photoreceptors in vertebrates (fish and land animals) but also in squid and octopus (marine invertebrates). Enlarged eyes gave great competitive advantages for feeding and escaping predation, and compound eyes conferred similar advantages on flying insects.

Human eyes closely resemble the eyes of octopuses. Both have camera vision. Camera vision does more than detect movement or coordinate movement, as among schools of fish, flocks of birds or herds of animals. Camera vision transmits a picture to the brain. The picture presents the brain with options for independent movement. Although human and octopus eyes are similar in appearance and use the same pigments to send images to the brain by electrical impulses, the nerve circuits are completely different. The transmitting nerves of human and octopus eyes evolved from different bundles of nerves during different geologic eras.

Why have two animals evolved two different nerve circuits for one of the most complex animal organs? The only plausible answer is random mutations that occurred in two different bundles of nerves in two very different animals, at two very different times, to serve the same needs. They were acquisition of food, escape from excessive predation and improved reproduction. If there is intelligent design, why were there two evolutions of highly complex camera eyes when one would have served both animals?

The story of human and octopus eyes is only part of the story. Eyes are so important for the survival of many animal groups that they have evolved at least forty different times along convergent pathways. One large group of animals has only eye-spots. Another large group, including humans, has large light-gathering camera eyes that aid the acquisition of food, shelter and finding mates. Other animals, like earthworms have no eyes. There is, thus, a continuous range of light-gathering organisms in animals, including eyes that detect color. Most of the evidence for the convergent evolution of eyes comes from investigation in comparative anatomy among living animals, not from the fossil record.

A second variety of convergence is body plans. Four unrelated orders of anteaters exemplify convergence. The four are: (1) the ant bear (edentate) of South America, (2) the aardvark of south and central Africa, (3) the pangolin (scaly anteater) of central Africa and southern Asia, (4) spiny anteaters (echidna) of Australia. Echidna belong to the earliest family of mammals because they have retained the reproductive anatomy of laying eggs after all other families of egg-laying mammals became extinct.

A third variety of convergence is flight. The geologic record indicates that flight has evolved at least four times among four very different groups of animals. The animals are insects, reptiles, birds and mammals (bats).

Insect flight originated in the Devonian era about 430 million years ago. Insects are arthropods that evolved atmospheric respiration of oxygen in order to feed on food resources in the tidal zone and further inland. Flight was a means of occupying new land niches where sufficient food was available to sustain reproduction. After the Cretaceous mass extinction, species of flying insects rapidly increased in numbers in conjunction with the rapid increase in species of flowering plants.

Flying reptiles evolved in the late Triassic (215 million years ago) and increased in size until the Cretaceous mass extinction. The largest flying reptiles had wingspans of fifteen meters. They probably ate fish. Smaller species are preserved in terrestrial sediments. They probably fed on insects.

Birds evolved from small feathered reptiles in the middle Jurassic (160 million years ago). Feathers are a form of scales and are very efficient in preserving body heat because they are hollow. Trapped air is a highly efficient insulator. Insulating feathers evolved into flight feathers as a means of escaping predation and accessing more and different foods. Because of their mobility they escaped extinction in the cretaceous mass extinction event. Bats evolved 53 million years ago (early Eocene) as insectivores. They accessed insects as food by feeding on them at night and in flight.

A fourth variety of convergence is echolocation. It occurs in two radically different animal groups: bats and toothed whales (dolphins). One uses sound waves in the atmosphere to locate food and the other uses sound waves in water to locate food.

A fifth variety of convergence is blind cave animals. Many unrelated animal groups retreated into caves to escape predation. Many species of fish, crayfish, salamanders, flatworms and insects occupy caves in widely separated regions of the world. Natural selection has operated in a uniform way on these animals. They lost pigment (became white) because pigmentation was not required in total darkness. They also became blind because eyes were useless in total darkness. In order to find food their senses of taste, smell and touch improved, with touch being aided by long feelers. Other evolutionary convergences were delayed reproduction until adequate food was available, larger eggs and fewer of them. Longevity also increased because reduced predation reduced pressure to reproduce at the youngest possible age.

SUMMARY

Diversity of life is propelled by genetic mutations that allow organisms to reproduce in new niches. The effects of mutations, however, are unpredictable because they are random. Natural selection operates only after mutations have occurred. Genetic mutations and natural selection are two unrelated biological processes. Genetic mutations are unpredictable but natural selection is highly specific for the single purpose of improving opportunities for reproduction.

Natural selection has preserved some bacteria without significant change for two billion years or longer. Other bacteria have evolved into more complex organisms. There is no linear evolution. There are no lower and higher organisms. Natural selection is an ongoing process that eliminates organisms that lose the ability to compete for food and preserves mutations that confer reproductive advantages. Natural selection does not produce perfection. It operates to produce survivors that can reproduce in immediately available niches.

Does the post-Cambrian expansion of complex life into many new niches mean there is direction to evolution? Humans like to think this is true because humans are, currently, at the top of the food chain. Humans flatter themselves with the simplistic belief that god directed humankind's elevated status in the biosphere, but this belief does not agree with genetics and the geologic record of life. The geologic record of life is a record of extinctions of most organisms in all geologic eras, especially animals at the top of the food chain.

Both bacteria and the complex organisms that are contemporaries of humans are equal products of evolution because they both reproduce in sufficient numbers to survive. Both simple organisms and complex organisms have equally survived the stresses of continuous changes in the biosphere. The only thing they have in common is the ability to reproduce.

CHAPTER 3

MASS EXTINCTIONS

Extinction is the death of species of organism without replacement. Mass extinctions are the sudden disappearance of large numbers of species without leaving descendants. The principal hard evidence is a sudden and precipitous decline in the variety of animals with visible shells or bones that are preferentially preserved in the geologic record. Mass extinctions of animals may or may not be accompanied by a sudden decline in the species of plants.

That said, what exactly is sudden, what is large numbers, what causes mass extinctions and how often do they occur? These four questions have variable answers because of gaps in the geologic record, but the occurrence of mass extinctions of animals is not in doubt. They are major events in the geologic record of life on earth.

Sudden presupposes a single catastrophic event or a cluster of catastrophic events. The best example of the suddenness of mass extinctions is the one that ended the Cretaceous era 65 million years ago. It was caused by a meteor impact. Within a month or two, dinosaurs became extinct worldwide.

The best answer to the second question is that specie losses vary from 50 to more than 90 percent, and family losses from 20 to 65 percent. The best example is the Permian mass extinction 251 million years ago. In excess of 90 percent of marine species became extinct, and the extinctions of land plants and animals were of the same magnitude.

The best answer to the third question is there are several known causes and several presumed causes. Mass extinctions may be caused by a single catastrophic event or a concert of catastrophic events. The best answer to the fourth question is there is no predictable time interval between mass extinctions. They are random.

FREQUENCY

The past 550 million years of the geologic record has preserved evidence of approximately two dozen mass extinctions. Some were large and others had less impact on the biosphere. The seven largest are: (1) 495 million years ago at the end of the Cambrian, (2) 438 million years ago at the end of the Ordovician, (3) 362 million years ago at the end of the Devonian, (4) 251 million years ago at the end of the Permian, (5) 206 million years ago at the end of the Triassic, (6) 65 million years ago at the end of the Cretaceous, (7) 40,000 years ago to the present.

The fossil record provides unambiguous evidence about the time and size of mass extinctions, but not about causes. Whatever the causes, the greatest impact is always at the top of the food chain. Overwhelmingly, the largest animals and plants become extinct. This opens the way for new organisms to evolve to fill vacant niches. After mass extinctions, the scope and direction of evolution is unpredictable.

CAUSES

This section will examine three mass extinctions: (1) disappearance of a high percentage of megamammals in Europe, central Asia, Australia and the Americas between 30,000 and 8,000 years ago; (2) disappearance of dinosaurs worldwide at the end of the Cretaceous 65 million years ago; (3) the mass extinction at the end of the Permian 251 million years ago.

Super Predation

Humans are the superpredators of megamammals. The extinction of a high percentage of megamammals in Europe, Asia, Australia and the Americas coincided with the migration of humans (*Homo sapiens*) from Africa to Europe, Asia and North and South America.

Homo sapiens entered Europe about 50,000 years ago and arrived in the Americas about 15,000 years ago. Within 7,000 years of their arrival in the Americas, paleontologists estimate that North and South America lost more than half of their megamammal species. Within 3,000 years of the arrival of humans in Australia (about 40,000 years ago), Australia lost 85 percent of its megamammal species (fifteen of sixteen genera of megamammals). On a smaller scale, the colonization of Hawaii about 1,500 years ago, Madagascar about 1,500 years ago and New Zealand about 900 years ago had a similar catastrophic impact on mega-animals and ground-nesting and flightless birds living on these islands.

Is superpredation by humans a past event? Clearly, no! It is a continuing event. Between 1870 and 1885, the herds of plains bison of North America were reduced from several million to a few thousands that survived in mountain valleys and in the northernmost forests beyond the agricultural frontier. Europeans hunted them nearly to extinction in order to practice agriculture, remove grazing competition for cattle, harvest their hides and force hunter-gatherer North American Indians to move onto reservations or starve. Bison would have become extinct but last-minute conservation programs ensured the temporary survival of a few small herds. (I use the term temporary survival because these herds could be exterminated in the blink of an eye if protection was withdrawn.)

One of the last extinctions of a megamammal was the Steller's sea cow. It was discovered in 1742 and hunted to extinction by 1769. The Steller's sea cow was the largest marine mammal other than whales. Similar predation in the nineteenth century nearly caused the extinction of sea otters and fur seals inhabiting the west coast of the North America. Many species of large whales came close to extinction in the 1970s because of worldwide human predation. The largest whale (blue whale) weighs more than 120 tons. It attains this size because the buoyancy of water supports its great bulk. Many other species weigh twenty tons or more. They migrate in herds and are easy targets for packs of small, fast killer ships operating in conjunction with factory ships. Concerted global conservation programs have enabled their temporary survival.

Humans used several migration routes out of Africa. The two preferred routes were north in the Nile Valley and island hopping at the south end of the Red Sea into modern Yemen and into southern Asia. Migrants who reached the Mediterranean Sea, at the mouth of the Nile River, went eastward to the Bosporus in Turkey and entered Europe via the Danube River Valley or migrated along the north shore of the Mediterranean Sea and went inland.

Other bands migrated westward along the south shore of the Mediterranean Sea until they reached the Straits of Gibraltar. Genetic evidence indicates that these migrating bands did not cross the Straits of Gibraltar. Instead, they migrated south along the Atlantic shore until they reached the arid zone of North Africa, and then went inland to where rainfall was more reliable.

Beyond the east end of the Mediterranean Sea humans used the shoreline of the Black Sea to enter central Asia and continue eastward. They migrated the length of Asia until they crossed into North America on the Bering Plain. They also migrated south and entered Australia by crossing the Arafura Plain. The Bering and Arafura Plains existed about 15,000 years ago because huge amounts of water were locked into continental glaciers during

MAP 1

Homo sapien Migration Routes Out of East Africa

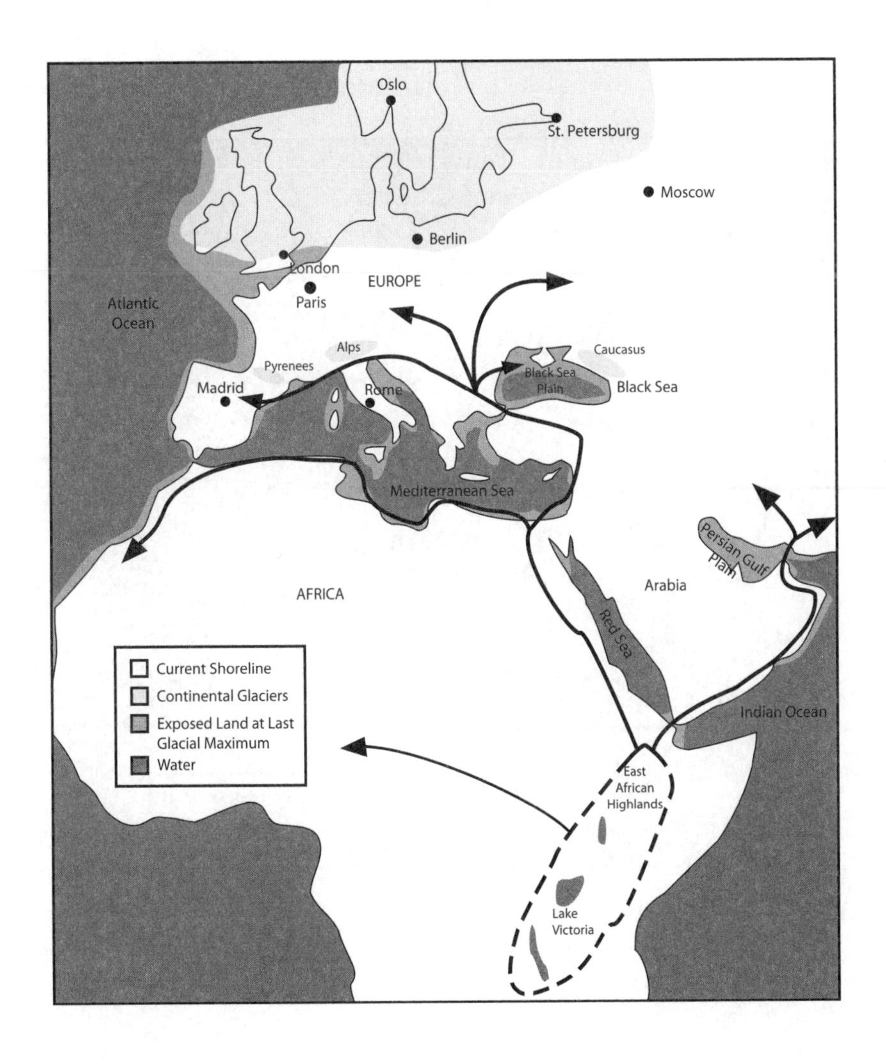

MAP 2

Exposed Land in Southeast Asia During the Last Glacial Maximum (21,000 years ago)

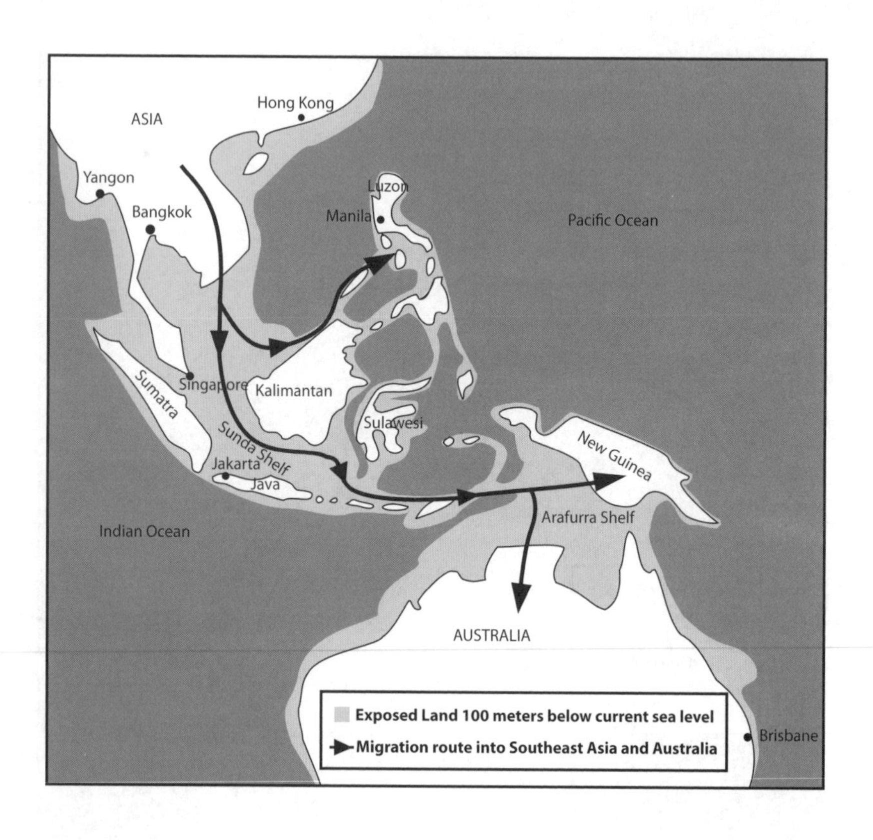

the last glacial maximum. The shorelines of all oceans were about 100 meters lower than present shorelines. Shorelines were preferred migration routes because coastal routes were warmer due to a thicker atmosphere, and food was abundant because it could be gathered from the tidal zone or caught from the sea.

After the continental glaciers of North America and Europe melted, the Bering and Arafura Plains were flooded. They became the Bering and Arafura Seas. The land links between Asia with North America and Asia and Australia were severed.

What food resources did *Homo sapiens* find when they entered the interior of Europe, central Asia, North America and Australia? They found themselves in the midst of a huge abundance of megamammals grazing on cool, temperate grasslands and browsing in cool, temperate forests. It was an unspoiled Eden of megamammals. In the interval between 30,000 and 8,000 years ago, a high percentage of the megamammals in Europe became extinct. This exactly coincides with *Homo sapiens* entering southern Europe and then migrating northward and eastward across central Asia and then south to Australia. Megamammals became extinct because they had little fear of humans.

In Europe, Asia and North America, extinctions of megamammals occurred along the north–south migratory routes of these animals. They were easy prey because they could be stalked. Humans were unfamiliar predators and they could closely approach them without causing a stampede. Spears were thrown and the hunters waited for the wounded animal to lie down and die. During their annual migrations north in the spring and south in the autumn, younger megamammals were preferentially hunted and the size of migrating herds steadily diminished—until migrating herds ceased to exist. Within a few generations, the only megamammals remaining were a few nonmigrating individuals in the sub-Arctic. These were also hunted to extinction.

In Europe, Asia and North America mammoths, woolly rhi-

MAP 3

Migration Routes into North America During Last Glacial Maximum

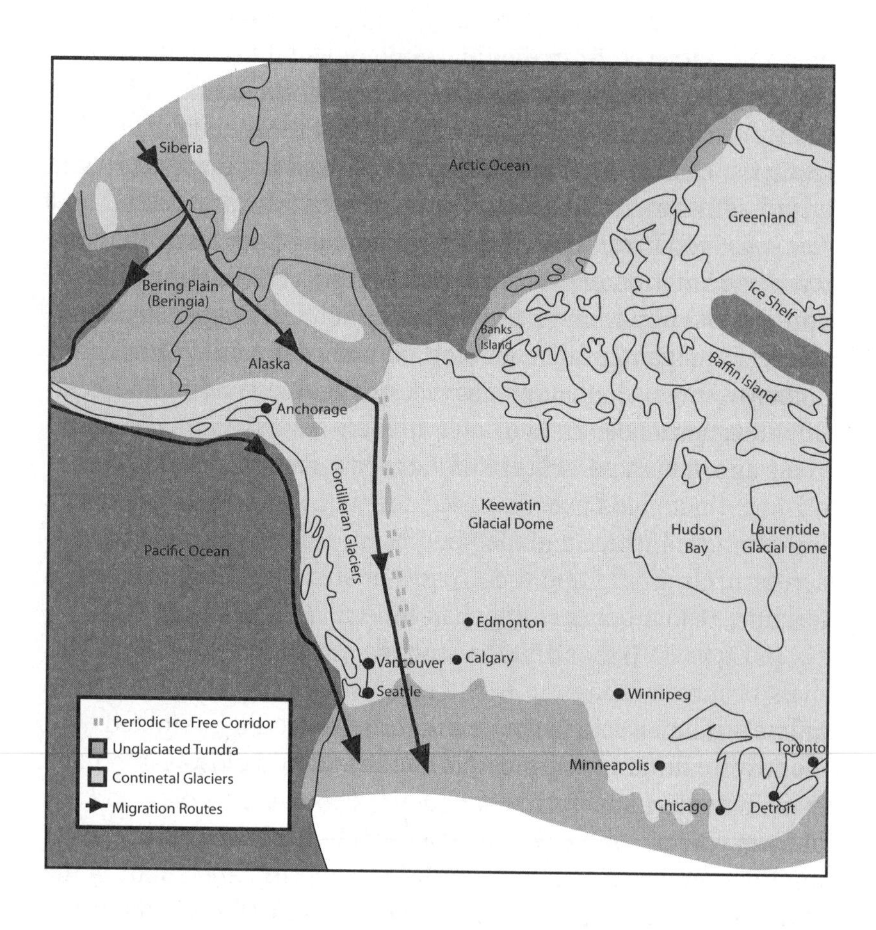

noceros, Irish elk and lions were hunted to extinction. Horses, camels, reindeer (European caribou), yaks in Asia and alpaca and llamas of South America survived because they were domesticated before wild herds were hunted to extinction. In Asia and North America only muskoxen survived because they could reproduce in the high Arctic where humans could not reproduce in sufficient numbers to kill all of them.

After humans arrived in America, about 15,000 years ago, the following megamammals became extinct: horses, camels, giant ground sloths, mammoths, mastodons, giant armadillo, glyptodonts, giant beaver, stag moose, long-horned giant bison, woodland muskoxen, short-faced bears, saber tooth cats and lions. In each of these groups, several to all of the species in them became extinct. Large numbers of smaller mammals (less than 44 kilograms) also became extinct. North America lost thirty-one genera of megamammals that weighed more than 44 kilograms. An exact replication of extinctions of megamammals took place in Australia after the arrival of humans.

In southern and central Europe, *Homo sapiens* clashed with a competitor for the high-protein diet of meat from megamammals. The competitor was the Neanderthal species of humans. The superior social organization of *Homo sapiens* in hunting and warfare exterminated Neanderthals because they were the principal competitor for the available supply of meat—the preferred source of nutrition for migrating bands of humans. Exterminating Neanderthal competitors was little different than exterminating lions in Europe, central Asia and central Alaska.

The only surviving megamammals in North America are species that had evolved in Europe and Asia and had already encountered humans before crossing the Bering Plain into North America. They had learned to fear humans. These animals were moose, elk, caribou, deer and bison.

Human migration to the Melanesian and Polynesian islands

had a similar exterminating impact on birds. Migration to Pacific islands began about 3,500 years ago. Since then, human predation coupled with the introduction of rats, has resulted in the extinction of at least 800 species, and the number may be more than 1,900. Many of the birds were flightless and others had no fear of man like contemporary penguins in Antarctica. The largest number of extinctions, however, was due to rats, because flightless birds with ground nests had no defense against nest predation.

Meteor Impacts

Meteor impact structures are preserved in all eras in the geologic record. Most of them are not obvious. As a byproduct of space exploration, particularly of the moon, geologists began a systematic search for impact craters on earth—and found them. As of 2009, more than 170 impact craters have been identified with diameters of one kilometer or larger. A high percentage were found by seismic and magnetic surveys. A geophysical signature is often the only evidence because most craters have been filled with sediments, eroded to their roots or completely covered by sedimentary rocks.

Meteor impacts on land have visible circular craters if they occurred in the relatively recent past. These structures are easily recognized because craters and rims are still intact. One of the best examples is Meteor Crater National Park in Arizona because it is in pristine condition. The oldest recognizable impact craters are about two billion years ago. These craters are on the order of 100 kilometers in diameter (or larger) because large size is necessary for preservation during long periods of erosion.

The moon's surface perfectly preserves meteor impacts because of the lack of atmosphere and liquid water. The number of large meteor impacts has been accurately counted, and there is every reason to believe that, relative to its surface area, earth has received as many meteor impacts as the moon. Most of these impacts, however, occurred in oceans.

All discovered impact structures on earth are on land because the basalt and overlying sediments that compose ocean floors do not preserve impact structures. Identified impact structures are probably about one-third of the actual impacts because more than 70 percent of the earth's surface is oceans, and ocean floors are continually subducted. The oldest ocean floor basalts are only 135 million years old. Ocean impact craters older than 135 million years have disappeared without a trace.

Unknown

Analysis of sediments deposited before and after mass extinctions indicates that five of twenty-four extinctions are associated with some evidence of meteor impacts. The strongest evidence is at the end of the Cretaceous era 65 million years ago. The second best evidence is at the end of the Devonian 362 million years ago—but evidence from the Devonian is far from certain because land plants were unaffected.

The geologic record indicates that oceans have covered about 70 percent of the earth for the past 600 million years. It is safe to assume that 70 percent of meteor impacts have been in oceans. Three-quarters of mass extinctions (nineteen) have only suggestive evidence or no evidence for meteor impacts. The cause of these extinctions can be interpreted as due to meteors hitting the ocean where no evidence of an impact is preserved. This is negative evidence and is, therefore, circumstantial. There is a continuing search for causes of mass extinctions in the geologic record, and meteor impacts are high on the list of possible causes. The search for causes, however, is still in its infancy except for the extinction at the end of Cretaceous.

CRETACEOUS EXTINCTION

There is a high degree of certainty that a meteor impact 65 million years ago caused the mass extinction that ended of the Cretaceous era. The Cretaceous extinction ended the tenure of dinosaurs and other megareptiles at the top of the food chain. The first evidence of a meteor impact was found in sediments that outcrop at several locations near Gubbio, Italy, a small mountain village north of Rome. At Gubbio, there are several continuous exposures of sedimentary rocks that span the end of the Cretaceous era and the beginning of the Tertiary era. These sediments define the K/T boundary. Within this section is a layer of clay one centimeter thick that is the exact boundary between Cretaceous and Tertiary sediments.

Cretaceous sediments below the clay layer contain many species of normal-sized forams and other carbonate-secreting microorganisms. Tertiary sediments above the layer contain only a few species of very small forams and the boundary layer itself contains no forams. Forams are single-cell marine organisms with carbonate shells. They live on the sea floor or float. Floating species are called plankton. Species of forams have been intensively studied because oil companies use their shells to define the ages of sediments that drills must penetrate to reach oil reservoirs.

It was a considerable accomplishment finding the precise location of K/T boundary sediments. The real significance of the layer, however, had to await laboratory analysis. This required asking the question: what can the centimeter-thick boundary layer tell us that cannot be seen in the field or with microscopes?

Chemical analysis of this clay layer in 1977 provided unambiguous evidence of a meteor impact. The evidence was highly anomalous amounts of iridium. Iridium is a rare element in the earth's crust. There are no iridium mines. Very small amounts of iridium are recovered when minerals containing platinum are refined; and platinum is a very rare element in the earth's crust. Iridium, however, is present in relatively large amounts in iron meteorites. A meteor impact was the presumed source of the iridium in the K/T boundary layer. Detection of the iridium anomaly, however, said nothing about the location of the impact.

The next question was obvious. Was the iridium anomaly at Gubbio caused by a local impact, or did sediments at the K/T boundary elsewhere have an iridium anomaly? If sediments at other outcrops that spanned the K/T boundary had an iridium anomaly, this would indicate that there had been a meteor impact large enough to affect all life on earth.

Locations of outcrops of sediments at the K/T boundary are reasonably well known. One of the most accessible is a coastal cliff in Denmark where the K/T boundary layer is thicker. The layer was sampled and it contained more anomalous iridium than the boundary layer at Gubbio. The discovery of the iridium anomaly in K/T boundary sediments was presented to the scientific community at an international meeting in 1979 and supporting data was published in 1980. Thereafter, K/T boundary sediments were analyzed worldwide. Within ten years, more than 100 outcrops of K/T boundary sediments were sampled. All of them had anomalous iridium in the boundary layer.

There is no evidence of a giant meteor impact crater anywhere near Gubbio, Italy, or near Denmark. The crater was discovered in 1988 along the north coast of the Yucatan peninsula of Mexico, near the coastal village of Chicxulub. Its discovery was due to a clear geophysical signature. The crater is now buried beneath about one kilometer of Tertiary sediments, with about half of it under land and the other half beneath the Gulf of Mexico. Concerted investigations indicated that the crater was large enough to affect the global biosphere and its impact was exactly at the K/T boundary.

The most important clue to its discovery was geophysical mapping done by PEMEX, the Mexican oil monopoly, in its search for additional domestic sources of oil. Seismic, magnetic and gravity surveys done on land outlined half of a giant impact structure but these surveys were buried in company archives, as were samples of rocks recovered from dry holes drilled in 1952. Holes drilled in 1952 penetrated about one kilometer of sediments overlying the geophysical anomaly, and then drill bits hit crystalline rock. The hole was stopped. These rocks were interpreted as volcanic. Oil is not found in volcanic rocks.

PEMEX geologists cannot be faulted for failing to recognize the origin of the crystalline rocks at the bottoms of dry holes. In 1952 there were only about eight people in the world who could have recognized that these were melt rocks caused by a meteor impact. PEMEX geologists did the next best thing. They preserved samples of the melt rock in a core library.

The meteor that ended the Cretaceous era is estimated to have been 10 kilometers in diameter. Its speed at impact is estimated to have been 30 kilometers per second. The crater it created was about 200 kilometers in diameter and 40 kilometers deep. Ejecta rocks, some of giant size, rained on surrounding land in northern Mexico, Cuba, Haiti and the southern United States, particularly in Texas.

Immediately after the dissipation of the energy released by the impact, the sides of the crater collapsed inward in giant landslides that produced numerous tsunamis as water sloshed back and forth in the confined water of the Gulf of Mexico. The initial tsunami is estimated to have been one kilometer high. It ripped up sediments from the continental shelf and carried a jumble of rocks far inland on the nearest coasts. Smaller particles were blasted into the atmosphere and dispersed worldwide.

The melt rock preserved by PEMEX was made available to academic geologists in 1991 so that analysis could be done in many laboratories. At the same time that melt rocks became available for analysis, additional K/T boundary sites around the rim of the Gulf of Mexico were being examined for glass spheroids. Glass spheroids are valuable because they preserve the original chemistry of the rocks melted by the impact. The texture of ejecta rocks that contain spheroids has been preserved in great details, but almost all of the glass produced by the impact has altered to clay. Only a few glass spheroids were preserved in surface sites, but unaltered spheroids were preserved in drill cuttings at the bottom of PEMEX's dry oil wells.

Spheroids discovered in K/T boundary sediments around the Gulf of Mexico and in PEMEX drill cuttings have the same geochemical signature. The elements in the glass corresponded to limestone and salt. The melt rock came from nonvolcanic rocks. The melt rocks in drill cuttings recovered from dry holes, that were preserved by PEMEX, were decisive in confirming that the Chicxulub geophysical anomaly was a giant meteor impact and that it was the probable cause for the mass extinction at the end of the Cretaceous era.

At the same time that the worldwide iridium anomaly was being confirmed and glass spheroids were being found and analyzed, other geologists were searching for additional evidence to confirm that a giant meteor impact caused the mass extinction at the end of Cretaceous. They knew what they were looking for. It was shocked quartz (a mineral named coesite) that is only formed by the intense pressures generated by meteor impacts. In 1983 they found shocked quartz in several outcrops of K/T boundary rocks. Shocked quartz indicated that the impact site was continental because quartz does not occur in basalt, the rock that forms the floors of oceans.

Another source of evidence is the ratio of carbon isotopes in boundary layer sediments. Photosynthetic marine microorganisms prefer to use the light isotope of carbon (carbon-12) to make their bodies, including shells. The shells of these organisms are enriched in carbon-12. Normal carbonaceous sediments are also enriched in carbon-12 because of their content of organic remains. In boundary layer sediments the two nonradioactive isotopes of carbon (carbon-12 and -13) accumulated in proportion to their abundance in the atmosphere. There was no enrichment of the carbon-12 isotope because the boundary layer contained very few living microorganisms. This is the carbon anomaly of boundary layer sediments.

The carbon anomaly is found worldwide. It indicates that: (1) there was a catastrophic reduction in the number of marine microorganisms when boundary layer sediments were deposited, (2) photosynthesis practically ceased because sediments that contain large amounts of organic carbon are enriched in carbon-12, (3) the food chain collapsed. The iridium anomaly, glass spheroids, shocked quartz and the carbon-12 anomaly in boundary sediments is extremely good evidence that a meteor impact caused the Cretaceous mass extinction.

The immediate effect of the impact was radiant heat from a fireball 300 kilometers in diameter that incinerated all organic matter for in a radius of 700 to 800 kilometers The worldwide cause of the extinction was a huge amount of dust and small particles blasted into the upper atmosphere at a speed of 5,000 kilometers per hour and dispersed worldwide. For many months after the impact a suffocating density of dust and small particles blotted out the sun.

Both oceanic water and land surfaces rapidly cooled. This was followed by an abrupt warming after dust and ejecta fell to earth. The warming was caused by the gasses that were generated by the huge amounts of sulfur dioxide, carbon dioxide and chlorine generated by the impact. The gasses were formed by the volatilization of limestone, gypsum and rock salt that were present in sediments at the impact site. In addition, the salt dissolved in marine water was volatilized. Rain converted the sulfur dioxide from gypsum into sulfuric acid, the chlorine from salt into hydrochloric acid and the carbonate from limestone into carbon dioxide gas. Both acids are toxic to marine and terrestrial life and carbon dioxide is a greenhouse gas.

The carbon dioxide created by volatilization of limestone remained in the atmosphere because there was greatly reduced numbers of photosynthetic plankton in the ocean and an equal decline in the number of land plants. Wildfires consumed much of terrestrial vegetation. Warming of marine water on continental shelves increased as much as 15 degrees centigrade. Land temperatures spiked higher. Both the cooling and warming of oceanic waters and land surfaces were sudden events and they happened globally. The abrupt reversals in the temperatures of oceanic water contributed to a precipitous decline in marine biomass and this decline persisted for 200,000 years or longer.

Between 60 and 70 percent of species of marine plankton and 70 percent of other marine organisms disappeared in sediments above the K/T boundary layer. Sediments immediately above the K/T boundary also had reduced carbonate deposition because a high percentage of carbonate-secreting microorganisms became extinct.

On land, the Cretaceous mass extinction was mostly an animal event. The most dramatic event of the meteor impact was the extinction of most reptiles. It was a dramatic event because so many species of reptiles were dinosaurs. These were the mega-animals at the top of the food chain. Paleontologists estimate that 50 percent of terrestrial animal groups became extinct, with megareptiles being annihilated. Insects, fish and flowering plants were less affected.

A few species of land animals in protected locations survived the catastrophic decline in food and the poisonous atmosphere. Many of the surviving species were small burrowing animals and a high percentage of other surviving species were marsupials. The principal marine mega-animals that survived were sharks, crocodiles and turtles. During the 10 million years after the Cretaceous mass extinction, to the beginning of the Eocene era (55 million years ago), mammals evolved to fill most of the land niches vacated by reptiles.

PERMIAN EXTINCTION

The Permian mass extinction was really two mass extinctions. One occurred 261 million years ago and the second ended the Permian era 251 million years ago. The extinction at the end of the Permian era is the largest in the geologic record. This is the extinction described in this section. Paleontologists estimate that around 95 percent of marine species and 80 percent of groups of terrestrial vertebrates became extinct. Terrestrial plant extinctions were of a similar magnitude.

The best evidence for the cause of the Permian mass extinction comes from sediments at the Permian–Triassic boundary. The evidence, however, is ambiguous. There are several locations worldwide where boundary sediments are continuously exposed without disconformities. The best outcrops are in South Africa and China, and the best studied outcrops are in the recently exposed walls of five quarries in China where phosphate rock is mined. Only a paleontologist could find the boundary layer.

The quarries are at Meishan, about 100 kilometers southwest of Shanghai. Like the K/T boundary layer exposed at Gubbio, the boundary layers exposed in the Meishan quarries are thin (30 centimeters). Sediments below the boundary layer contain abundant fossils of late Permian marine microorganisms. Above the boundary layer over 90 percent of species of carbonate secreting microorganisms disappeared and, like the Cretaceous mass extinction, there was a precipitous decline in marine biomass.

Within the boundary layer at Meishan is a thin bed of black carbonaceous shale. Like the carbonaceous shale beds at the at the K/T boundary at Gubbio, Italy, the carbonaceous layer contains the same percentage of carbon-12 isotopes as the atmosphere because marine micro-organisms were not there to use the carbon-12. Most of the microorganisms that used carbon dioxide to make shell were extinct. Photosynthesis practically ceased. This strongly indicates that the food chain collapsed.

There are two principal disagreements among geologists. How quickly did the Permian mass extinction occur? What caused the extinction? The best answer to the first question is that the onset of extinction appears to have been a sudden event and it occurred in the oceans and on land. From a distance of 251 million years, sudden can mean 100,000 years, or less than 10,000 years, or less than 100 years. Suddenness would indicate that the triggering event was a meteor impact or a series of closely spaced impacts that kept the atmosphere filled with dust, micro-ejecta, carbon dioxide and other poisonous substances.

Alternatively, the extinction could be due to the conjunction of a meteor impact with another stressful event. The best candidate for a coincident event is the huge outpouring of flood basalts in Siberia that coincides with the end of the Permian era. Flood basalts are common events in the geologic record but the Siberian event is one of the largest. It is estimated to have erupted more than four million cubic kilometers of basalt that covered seven million square kilometers.

Eruptions of flood basalt are not explosive. They are generally quiet events that continue for substantial periods of time. Most basalt magma reaches the surface through closely spaced cracks in the crust. These cracks are preserved as diabase dikes. Diabase dikes are composed of basalt that did not reach the surface as lava flows. The basalt in the cracks froze in place. The largest dikes are often more than 100 meters wide and longer than 100 kilometers. Conduits this size had the ability to carry large volumes of lava to the surface.

Single Siberian lava flows can be more than 300 meters thick but they average about 80 meters. Flood basalts create thick piles of stacked flows. In some places in Siberia, the pile is 6,500 thick. As lava flowed onto the surface, the crust subsided on top of the magma chamber where the basalt originated.

Was the outpouring of Siberian flood basalts the only cause of the Permian mass extinction? Or was it a contributory cause? Making it the only cause is a very attractive explanation because of the extent and visibility of the flows. It is also an attractive explanation because the radiometric age of some of the flows corresponds with the radiometric age of volcanic ash found in sedimentary rocks in Permian–Triassic boundary layer in China.

A volcanic cause of the mass extinction is attributed to the release of large quantities of carbon dioxide, sulfur dioxide and methane into the atmosphere. These are greenhouse gasses. A large volume of methane is assumed to have originated from heating the largest coal basin on earth that underlies the pile of Siberian flood basalts. Gasses released from the coal were dissolved in the basalt magma, and when the magma reached the surface the gasses were vented into the atmosphere. The sulfur dioxide combined with water vapor to form sulfuric acid that had a devastating effect on terrestrial vegetation.

Although gasses released from basalt flows probably contributed to fluctuations in the temperature of the atmosphere, it is unlikely that there was a sufficiently large surge for a sufficient number of years to cause a mass extinction. The outpouring of Siberian flood basalts continued for least 300,000 years, and probably much longer. But, a volcanic cause of the Permian mass extinction cannot be ruled out.

If, however, a meteor impact coincided with a major outpouring of basalt or an impact triggered a major outpouring of lava, the effects would be catastrophic. A worldwide temperature decline would result from dust and micro-ejecta blotting out sunlight, followed by a temperature rise induced by a huge addition of greenhouse gasses to the atmosphere from the impact and from volcanism. The marine and terrestrial food chains would collapse.

Although a meteor impact (or several nearly simultaneous impacts) is the best candidate to explain the Permian extinction, the evidence for it is ambiguous and difficult to interpret. (1) Boundary layers at several Chinese localities lack anomalous iridium, but the dust and micro-ejecta from a stony meteorite would not have had a sufficient amount of iron/iridium alloy to cause an anomaly. (2) At several Chinese localities and one in Antarctica, microparticles of an iron-nickel alloy mixed with particles of silicate minerals have been recovered from the boundary layer. (3) Glass spheroids are found in boundary layer sediments but not shocked quartz. (4) Shocked quartz would not be present in boundary layers if a meteor impacted in the ocean because quartz is absent from the basalts that compose ocean floors.

Finally, there is fragmentary evidence (2009) of a giant impact structure in the Antarctic, but confirmation is in its infancy. Geophysicists working in Antarctica believe they have discovered an impact structure larger than Chicxulub. It is buried under several kilometers of ice. Preliminary measurements of its geophysical signature indicate the structure is about 500 kilometers in diameter. A crater this large would require a meteor that was at least 20 kilometers in diameter. The structure outlined by the geophysical anomaly has not been tested by drilling to confirm that it really is a crater. And if it is a crater, what is its age? If it is an impact structure, based on evidence from the Chicxulub structure, it would probably be large enough, in conjunction with the outpourings of Siberian flood basalts, to cause the Permian mass extinction.

Triassic Recovery

Reptiles were a high percentage of surviving terrestrial animals. The largest was about one meter long. Other survivors were several species of reptiles that had mammalian characteristics but were not quite mammals. In the late Permian some premammalian species were as long as two meters but only mouse-size mammal-like reptiles survived into the Triassic. Mammal-like reptiles lost the evolutionary race to dominate post-Permian terrestrial life because a larger number of large reptile species survived. Size was highly advantageous in the competitive race to fill huge numbers of vacant niches available after the mass extinction. Reptiles won the feeding and reproduction race and dominated the 186 million years of the Mesozoic era.

Full recovery from the Permian extinction required more than 10 million years. The early Triassic ocean and land was populated with huge numbers of the few species that survived the extinction. Evolution to fill old terrestrial and marine niches was slow because of the magnitude of the Permian extinction. Many old niches were empty until new species evolved to fill them. As new species evolved, many survival species retreated into their pre-extinction niches and became minor fauna in the late Triassic fossil record. The niches for large herbivores and carnivores were the last to be filled.

At the same time that reptiles evolved to dominate terrestrial and marine habitats, mammal-like reptiles evolved into true mammals. Warm blood made it possible for them to become nocturnal feeders because they retained a high energy level during the cool of the night. They were agile and inconspicuous and slowly increased their range because they could expand into cool climates where cold-blooded reptiles were at a disadvantage.

Tertiary Recovery

The Cretaceous mass extinction propelled a comprehensive restructuring of life in the Tertiary era. There was an evolutionary free-for-all in body plans on all continents. The extinctions of dinosaurs meant that almost all terrestrial niches for mega-animals were empty. Mammals up to the size of house cats were the preferential survivors of the Cretaceous mass extinction. The two principal families of surviving mammals were marsupial and placental. Marsupials were more widely distributed.

The Australian Plate rifted from the supercontinent of Pangea about 225 million years ago. Marsupials had evolved before Pangea rifted from the supercontinent but placental mammals had not yet evolved. In Australia marsupials filled most of the niches vacated by megareptiles. Marsupials were less active competitors for food acquisition than placentals, but they survived as dominant animals in Australia because there were no placental competitors. Lack of placental competitors allowed bizarre mammals to survive. They are the anteater echidna and the aquatic platypus. Both are egg-laying mammals.

Like Australia, South America was a giant island that was not attached to another continent. During the 25 million years after the Cretaceous mass extinction, marsupials evolved in South America and Australia to fill most niches. In South America a large family of flightless carnivorous birds filled the vacant niches of foxes, wolves and lions. The largest was 3 meters tall (Titanis) with a hooked eagle beak 38 centimeters long.

Flightless carnivorous birds and marsupial carnivores became extinct less than three million years ago after subduction tectonics created the Panama land bridge between North and South America.

South America was invaded by placental carnivores from the north. Foremost among them were leopards (jaguars). Within several hundred thousand years all species of flightless carnivorous birds were extinct, as were most marsupials. Placental carnivores replaced carnivorous marsupial and flightless birds at the top of the food chain. Replacement of marsupial by placental mammals had previously happened in Eurasia and North America.

During the 10 million years after the Cretaceous mass extinction (65–55 million years) reproduction favored placental herbivores that could eat grass. Grasses are post-Cretaceous plants. The earliest fossil pollen of grass are about 40 million years old. By 30 million years ago, grass covered large areas of North America, Africa and Asia that had highly seasonal rainfall. Grazing animals (herbivores) evolved to feed on them. They evolved hooves for speed to escape predation, chisel teeth at the front of jaws to cut grass blades, flat teeth at the back of jaws to grind them and multiple stomachs that contained symbiotic bacteria that could metabolize the high cellulose content of grass into sugars that were then metabolized by herbivores.

Parallel to the evolution of large grass eating herbivores were large social carnivores (wolves and lions) that preyed on them. As near as the geologic record can tell us, the mammalian herbivores and carnivores that evolved in the Tertiary era were an exact convergence with herbivorous and carnivorous reptiles that had evolved during the Mesozoic era. Put in other words, the niches occupied megamammals in the Tertiary were highly similar to the niches occupied by megareptiles in the Mesozoic era.

Recent Recovery

The extinction of megamammals during the past 30,000 years is too recent for evolution to replace them. It is much more likely that human predation (that began 30,000 years ago) will continue to reduce biodiversity in animals of all sizes. The mass extinction of near time is not yet complete.

SUMMARY

Paleontology documents many mass extinctions that have occurred at irregular intervals in geologic time. Paleontology also documents that extinctions of animals and plants are ongoing events. The causes of most mass extinctions can only be surmised because only limited evidence survives from 100 to 500 million years before the present. There is no doubt, however, that mass extinctions occurred. What are probable causes for pre-Cretaceous mass extinctions? Catastrophic events like meteor impacts are a favorite explanation because many large impact craters are preserved in the geologic record. Another proposed cause is surges of greenhouse and poisonous gases from major flood basalt events. The geologic record preserves many flood basalt events of varying size, and some of them correlate with mass extinctions. Another proposed cause is continental collisions that raised mountain ranges, like the Himalayas, that radically altered rainfall patterns over continental-sized areas.

These collisions would also reduce areas of continental shelves where reproduction of marine animals is concentrated. Two mass extinctions have clearly documented causes. The first is the meteor impact that ended the Cretaceous era 65 million years ago and was responsible for the extinction of most terrestrial and marine reptiles (dinosaurs). The second mass extinction is ongoing. It is the super predation of megamammals by humans (*Homo sapiens*). It began 40,000 years ago when humans arrived in Europe, Asia, Australia, and the Americas. In geologic time, human extermination of megamammals in the last 40,000 years is an instant event.

CHAPTER 4

ORIGINS OF AMERICAN FUNDAMENTALISM

American fundamentalism began in the Burned Over District in northern New York in the 1830s. Mormonism was one of the first sects to form in the search for new revelations suitable to the United States as the world's first nation with democratic governance. In 1830 Joseph Smith published the *Book of Mormon*. He claimed that the angel Maroni revealed to him that his disciples would evangelize the world. The *Book of Mormon* claims that the new Jerusalem would be in the United States, and this is where the second coming of Christ will occur.

Smith was assassinated in 1844, but Brigham Young, a brilliant organizer, guided a migration of true believers to Utah where they settled on the banks of a river they named Jordan, and along the shores of Lake Utah (Sea of Galilee), and along the shore of the Great Salt Lake (Dead Sea). In their imagination they believed they were the revived nation of Israel with Salt Lake City as the new Jerusalem.

Several other sects, based on millennial expectations, became part of the fundamentalist movement. Seventh Day Adventist is based on revelations received by William Miller, a farmer in northern New York. He preached the second coming of Christ in near-time. He set a date in 1843 and then in 1844. When there was no visible return, Miller's associates claimed Christ returned in spirit and was, therefore, invisible. His physical coming would be in the indefinite future.

A third millennial sect was Jehovah's Witnesses. It had its origins in revelations experienced by Charles Russell. In 1874 he proclaimed that Christ had made an invisible return to earth in 1874 in preparation for his visible return in 1914. Instead of Christ's second coming in 1914, the world got World War I. Predictions continued to be made—1918, 1925, 1974, 1984 and 1994. True believers continue to wait.

Pentecostal Assemblies of God originated about 1910 in California, probably in churches in the black community. It was a religion of illiterate or semiliterate peasant migrants (black and white) from the cotton south who were evicted from land they cultivated at a subsistence level of livelihood. They were forced to migrate to an urban commercial culture, for which they were unprepared. They worked as agricultural and urban laborers.

Pentecostal gets its name from the day of Pentecost. In the Christian calendar it is seven Sundays after Easter. It celebrates the descent of the Holy Spirit on a gathering of apostles who were expecting the imminent return of Jesus. The gathering was a frenzy of expectations that the second coming would occur in near-time. All of those present babbled unintelligible sounds (glossolalia) until the frenzy ended in a trance.

> When the day of Pentecost had come, they were all together in one place. And suddenly a sound came from heaven like the rush of a mighty wind, and it filled all the house where they were sitting. And there appeared to them tongues as of fire, distributed and resting on each one of them. And they were all filled with the Holy Spirit and began to speak in other tongues, as the Spirit gave them utterances. [Peter then explained the meaning of their experience.] Repent and be baptized every one of you in the name of Jesus Christ for the forgiveness of your sins; and you shall receive the gift of the Holy Spirit. (Acts 2: 1–4; 38–39)

Religious services of Pentecostal congregations often duplicate the frenzy of direct spiritual contact with god and Christ on the first Pentecost. Pentecostal frenzies usually occur in small congregations composed of persons seeking certitude for their salvation. The best certitude is a personal experience. Three of the most common experiences are a trance, babbling and uncontrolled weeping for the joy of salvation. Exhortations to induce these behaviors are especially effective on young persons at an impressionable age.

Other prophetic denominations that appealed to poorly educated persons were: Church of God (1881), Christian Missionary Alliance (1887), Church of the Nazarene (1895) and Church of God in Christ (1897). Initial leaders of these denominations were marginally literate clergy who believed that the experience of being baptized by the Holy Spirit was an experience that assured salvation. In the United States, true believers are highly concentrated in Pentecostals, Churches of God and Southern Baptists, plus many smaller sects.

AMERICAN EXCEPTIONALISM

The United States is not like other nations. God has directed it to evangelize the world. This is a two-step process in the contemporary world. As the first nation to practice democratic governance, the United States has a mission to induce democratic governance in the other nations of the world. Secondly, as the only superpower in the world, the United States has been chosen by god to cleanse the world of evil empires. This is a necessary step to evangelize the world in preparation for the second coming of Christ. Only super power exercised by the United States can prepare the world for global evangelism and the second coming of Christ.

CONTEMPORARY FUNDALMENTALISM

In 2006 fundamentalists composed about 25 percent of church membership in the United States. The largest percentage is Southern Baptists, with various sects of Pentecostals next. Approximately 75 percent of persons in these denominations expect the apocalypse to occur in near-time.

Fundamentalist clergy prophesy that the apocalypse will exterminate all evil empires. The apocalypse is the global war that will be the occasion for Christ to return to earth and lead the armies of righteousness to victory over the armies of evil empires. This victory will precede the end to time. True believing Christians feel that the most menacing evil empires are Muslim because the last remnant of the evil empire of communism is North Korea.

When Christ returns to earth he will raise the souls of true believing Christians from the dead, reinsert them in their physical bodies and rapture them into heaven where they will enjoy eternal life. Belief in the apocalypse and rapture has an unintended result. True believers will be in heaven and the populations of evil empires will be exterminated or in hell. Humans will become extinct on earth.

HISTORICAL ANALYSIS

Fundamentalist clergy reject historical analysis of biblical texts. They reject explanations of the historical and political conditions that motivated prophets and apostles to write texts that clergy in the post-apostolic age judged sufficiently important to be included in the bible. Fundamentalist clergy reject all textual analysis because they want no modification of the literal truth of every event in the bible that they believe is the inerrant word of god.

Fundamentalist clergy mock "graduate-level words" like exegesis, eschatology, anthropomorphic, Christology (Strandberg, *Are You Rapture Ready?*, 90) because theologians use these terms to encase the simple gospel message of salvation into doctrines and dogmas that do not proclaim the message of faith and the saving power of the Holy Spirit. Fundamentalist clergy believe that critical analysis of biblical texts is irrelevant—or more likely evil—because all prophetic and apostolic messages are literally true. They believe the original messages that were addressed to hugely different people, living in radically different circumstances, have a simple validity that crosses time, cultures and circumstances.

Fundamentalist clergy are taught to preach the message of repentance and salvation by using snippets of text extracted from anywhere in the bible without relation to their historical context. Curriculums of bible colleges ignore the historical circumstances that motivated the authors to write the books included in the bible because historical analysis questions the inerrancy of biblical texts.

Almost always, the snippets of texts that fundamentalist clergy extract from the bible have little relationship to the current problem they address. For example, both the Old and New Testaments assume that the organization of family life and congregational governance was patriarchal. Men had high social and political status and women had a low social status. Two of the best measures of woman's low status are the prevalence of polygamy in the Old Testament and the apostle Paul's instruction that women remain silent in congregational worship (1 Corinthians 14:35).

The best example of radically changed circumstances is the status of slavery. In the Old and New Testaments it was an accepted institution. Neither the prophets of the Old Testament, nor Jesus, nor the apostles condemned it. Paul said slaves should obey their masters (Colossians 3:22; Titus 2:9). In the United States, from 1830 to 1860, most clergy in slave states defended slavery because it was a god-sanctioned institution. Almost all southern clergy supported the Confederacy because they considered slavery essential for the preservation of a patriarchal society that was governed by white planters who were benevolent protectors of the welfare of their black slaves.

In the democratic and industrial cultures of the United States and Europe, slavery, polygamy, racial segregation and patriarchal family life are no longer normative. They are either illegal, immoral or both. Does the inerrant word of god in an inerrant bible continue to validate these practices?

There are other serious problems with literalism. Jesus was an oral communicator of great presence who had an enormous impact on some listeners. His life was a teaching mission but he taught in Aramaic. The gospels, however, are written in Greek. The language changed between Jesus' teaching and when the gospels were written.

In addition, an indefinite number of years passed from the time Jesus taught to the time when written accounts of his lessons were assembled in the gospels. Some listeners wrote accounts of his lessons soon after they were delivered, and these accounts had a wide circulation among congregations started by the apostles. When these lessons were assembled into the gospels at a later date their primary purpose had changed from attracting Jewish believers to attracting gentile believers.

POLICY IMPACT

How do fundamentalist beliefs impact policies of governance in the United States? Fundamentalist clergy strongly support three policies: (1) bias against funding scientific research (except for medicine, agriculture and armaments), (2) prohibiting abortions, (3) uncritical support of the nation of Israel.

How do fundamentalist Christians seek to reduce the impact of science in American culture? They actively campaign to elect true believers to state and local school boards. These boards adopt policies that minimize teaching science by controlling the funding of primary and secondary education and controlling curriculums. Alternatively, the minimal teaching of science can be achieved by adopting textbooks that exclude teaching many life sciences, especially evolution.

At the university level, fundamentalist clergy distrust public funding of research in many scientific disciplines. They especially oppose public funding to investigate many problems in the life sciences because the hypotheses that frame many of these investigations lack a god content.

Are fundamentalist clergy antiscience? Yes, but with qualifications. They recognize that science has been responsible for ending famines and plagues in commercial cultures. They also want a prosperous commercial culture that funds continuous scientific research in desirable disciplines. The most desirable disciplines are medicine, agriculture and armaments. They also want to reduce the scope of scientific research in disciplines they consider undesirable because they believe they threaten true belief. This is especially true of the evolutionary sciences of paleontology, genetics and molecular biology that strongly confirm that natural selection has governed the evolution of life from life's inception.

ABORTION

The claim that life begins at conception has no basis in science. A fetus is formed at conception, but life does not begin until a fetus is expelled from a female body and can breathe air. Spontaneous abortions are frequent after conception. Fundamentalists and Roman Catholics make no distinction between spontaneous abortions (usually caused by genetic mistakes or malnutrition) and induced abortions (usually for economic and medical reasons). According to clergymen, all abortions are evil because a soul is lost. Nothing is said about preserving the health of women.

Abortion is a public health service. Pregnant women who are addicted to heroin, cocaine, alcohol or who have AIDS should be led to abortion clinics because their children will be damaged from birth. Many or most of the children born to these women will become burdens on the public revenue for large parts or all of their lives. They will become burdens because they will be unable to work or unable to keep a job for any length of time. There is also considerable evidence that a high percentage of them will be mentally disturbed and inclined to crime because they have been raised in nonfunctional households.

Teenage girls who become pregnant should also be led to abortion clinics because many or most of them will not remain in school long enough to acquire functional literacy. Functional literacy is the fundamental skill in commercial cultures. Their illiteracy will program most of them to live on welfare for some time during their lifetime.

Life is not sacred as claimed by fundamentalist and Roman Catholic clergy. If life were sacred, Christian clergy would be against all wars. Fundamentalists do not compare the termination of damaged fetuses to the deaths of young men in wars when they are in the prime of their lives. War is approved social conduct, abortion is not. War deaths on a huge scale are acceptable, but termination of genetically damaged or drug-damaged fetuses is not.

Fundamentalist and Roman Catholic Christians easily accept violent death because violent death pervades the Old Testament.

There are concise accounts of internecine warfare approved by god in the books of Numbers and Judges. In Judges there is account of how a gang rape by homosexuals was averted by substituting a concubine for the man. She died from the ordeal. This event initiated tribal warfare in which thousands of men were slain. Many villages were captured and their inhabitants massacred. Only young women were saved. They were spoils of war who were divided among winning warriors to be wives or concubines (Judges 19:22; 20, 21; Numbers 31:1–18).

Christian clergy have supported many wars in the historical past, just as god in many Old Testament narratives supported military aggression. Wars supported by Christians are often called crusades. The apocalypse predicted by fundamentalist clergy is a holy war of extermination against nonbelievers.

PRO-ISRAEL FOREIGN POLICY

Fundamentalist Christians believe that the apocalypse that precedes the second coming of Christ will be fought at Armageddon (Meggido). Meggido is an Old Testament battlefield site within the nation of Israel; therefore, the nation of Israel must be preserved from Muslim conquest because this is where Jesus will return. This is confirmed by many texts in the Old Testament that refer to Jerusalem as the city of god. "In the mind of God Jerusalem is the center of the universe." For most fundamentalists, preserving Israel and Israel's control of Jerusalem is a necessary preparation for the second coming of Christ because "We are racing toward the end of the age. Messiah is coming much sooner than you think!" (John Hagee, *Jerusalem Countdown*, 96, 99) When Jesus returns, "All Israel will be saved" (Romans 11:26). He will convert the Jews who failed to convert in his lifetime and the rapture will end time.

SUMMARY

The fantasy world of fundamentalism is complete.

1. The creation myth in Genesis is a historical fact.
2. The United States is god's chosen instrument to evangelize the world in preparation for the second coming of Christ.
3. In the near future the apocalypse will occur.
4. During the apocalypse Christ will return to earth to lead an army of true believing warriors.
5. He will preside over the rapture into heaven of living and dead true believers.
6. Nonbelievers will be consigned to hell.
7. Humans will become extinct on earth.
8. Time will end.

CHAPTER 5

INTELLIGENT DESIGN

Intelligent design is a concept that claims that life could not have originated by a series of spontaneous chemical reactions, and that natural selection is not capable of evolving the complexities of the biosphere. Believers in intelligent design claim that an omnipotent god created the biosphere and is its governor, and humans are god's surrogates to govern the earth's biosphere. This claim has three parts: (1) origin of life, (2) complexity of the biosphere, (3) humankind's status in the biosphere.

CREATIONISM

A high percentage of believers in intelligent design believe that the Genesis myth describes the actual origin of life and that intelligent design confirms its validity.

Origin

True believers claim that life on earth was created by a supreme being (god) who transcends space, time and physical matter. They claim that the creation of humankind was a special act by an omnipotent god in order for humans to govern the biosphere for their benefit. In other words, the Genesis myth puts the earth and humans who inhabit it at the center of life in the cosmos.

Complexity

Believers in intelligent design claim that the complexity of genetics and all other biological sciences could not have originated by physical processes alone. It is clearly impossible that evolution based on natural selection has always governed the biosphere because natural selection cannot explain the biological complexities of all living organisms. Biological complexity requires a designer. This assertion is not based on any science. It is based on the revelation that god communicated to Jewish patriarchs long before science was invented.

Revelation

Believers in intelligent design reject evolution by natural selection because it is a wholly physical process. It contradicts the power of an omnipotent god, revealed in the Genesis myth, that humans are a special creation of god for the special purpose of governing the biosphere. Revelation must overrule all scientific evidence that diminishes belief in the existence of an omnipotent god because humans are not accidents of evolution. They are a purposeful creation of god.

FUNDAMENTALIST CHRISTIANS

Fundamentalist Christians are believers in the literal truth of Holy Scripture. They believe that Scripture is infallible because the truths in it were revealed to god's chosen servants from the time of Abraham to the time of Jesus and his apostles. They believe that the greatest challenge to the literal truth of Holy Scripture is science, particularly geology and the related sciences of genetics, molecular biology, evolutionary biology and paleontology. All of these disciplines presume indifference to god's existence or actively reject god's existence because science substitutes natural selection for faith in the omnipotence of god.

How do fundamentalist Christians avoid the overwhelming scientific evidence for the operation of evolution by natural selection? They do it the simple way. They deny its existence. Believers in the truth of the Genesis myth do not require any knowledge of science, nor do they require literacy. The great strength of the Genesis myth is that it is a simple explanation for humankind's status at the top of the food chain.

The Genesis message is highly acceptable to these people because it elevates humankind into a special status with god. This status is available to all persons who accept the Genesis account of creation as god's revelation to humankind. Intelligent design is a logical deduction from the Genesis account of creation. Persons who accept the certitude of the Genesis myth are true believers.

There are two categories of fundamentalist Christians who reject natural selection. They are young earth and old earth creationists. Young earth creationists believe that it is literally true that god created the biosphere as it now exists in six days about 10,000 years ago. The culmination of god's creative power was the separate creation of humans (Adam and Eve) in order for humans to fulfill god's plan that they govern the biosphere.

Old earth creationists accept the concept of evolution after god created life 4 billion years ago. For them, evolution is a many step process over a long period of time that god has actively guided in order to create humankind to govern the biosphere. Both young and old earth creationists reject natural selection as the propellant of evolution.

But why do fundamentalists accept two versions for the creation of humans (Genesis 1: 26–27; Genesis 2: 7–8, 18–22)? If fundamentalists believe that the account of creation in Genesis is a historical event, surely, one creation of humans is adequate?

SCIENCE

All science has four components: (1) observation, (2) hypothesis, (3) experimentation, (4) replication. If any of the first three components are flawed, the result is an experiment that will not replicate. If an experiment will not replicate, it is not science.

Science is a search for causes. It begins with accurate observation of a phenomenon followed by questions of how or why the phenomenon occurs. Testing for causes cannot begin until there is a hypothesis about a probable cause and this hypothesis must be capable of being tested by experiments. Alternatively, science accumulates observations that fit into a coherent hypothesis that can reasonably explain many related phenomena. Evolution by natural selection is a hypothesis that does this by integrating observations and experiments in geology, paleontology, taxonomy, stratigraphy, genetics, molecular biology and all the other life sciences.

Scientific experiments are designed to confirm or reject hypotheses. A hypothesis is not valid until multiple experiments can replicate it, or an array of observation can be organized into a coherent explanation that is supported by evidence from related scientific disciplines. If experiments confirm the validity of a hypothesis, scientists invite other scientists to perform the same experiment in order to replicate results. If the results are replicated, it confirms the validity of the hypothesis. The hypothesis then becomes a guide for more experiments by more scientists in order to increase the certainty of the hypothesis and expand its scope.

If an experiment cannot be replicated, the hypothesis must be reformulated or abandoned. Some experiments have ambiguous results. Ambiguous results are neither right nor wrong. They are inconclusive and await further experimentation until the hypothesis that was used to frame it is modified or rejected. In the process of continual experimentation some hypotheses acquire a high degree of certainty; others are rejected; and others remain ambiguous until new observations are made or new instruments are used to produce conclusive results. Science is open ended in order to continually include more experimental data that increases the scope of understanding. This is inductive reasoning.

PALEONTOLOGY

Paleontology is the study of fossils preserved in sedimentary rocks. Shells and bones are the most frequent animal fossils. Under exceptional circumstances soft tissue is preserved. In a few places on earth they are preserved in exquisite detail. Mostly, however, paleontology is the comparative study of shells and bones. The first animals with shells entered the geologic record in the latest Vendian era 555 million years ago. They were only visible under a microscope but, for the first time biomineralization became a visible part of the geologic record.

Shells, exoskeletons and bones define body plans. In the lowermost Cambrian sediments shells and exoskeletons were one millimeter or less in size. In many cases it is not clear whether biomineralized fossils were carapaces for a single animal or overlapping scales on a soft body. Shells and exoskeletons continually increased in size during the rest of the Cambrian era and evolved into highly variable forms with highly variable chemical compositions.

Arthropods (trilobites) had chitin exoskeletons; bivalve animals (clams) used calcite (calcium carbonate) to make shells; a few animals used apatite (calcium phosphate) for scales and shells; and others (sponges) used silica (silicon dioxide) to make spicules to strengthen soft tissue. By the middle Cambrian, shells and exoskeletons encased many animal species. Paleontology is the best measure of the scope of evolution because it is highly visible. Strong supporting evidence for the timing of the evolution of animals with shells comes from the related disciplines of molecular biology, genetics, stratigraphy and geochronology.

GEOCHRONOLOGY

Geochronology measures the ages of rocks that compose the crust of the earth. The geologic age of rocks is determined by the decay of radioactive elements in minerals that compose crustal rocks. Radioactivity transmutes one element into another. The rate of decay of radioactive elements in minerals that compose crustal rocks measures the length of geologic time since that rock crystallized from a magma. Radioactive age-dating depends on a physical constant—the rate of decay of radioactive elements. Decay rates are constant in uranium, potassium, carbon and all other radioactive elements used in age-dating. Decay is unaffected by atmospheric temperatures, pressure or any other known physical or chemical reactions. There is no subjectivity.

The constant rate of decay can be calculated, and it is usually expressed in half-lives because of the long time intervals for radioactive decay to go to completion. Calculating the half-life of radioactive elements is done by measuring the ratio of end product elements in a molecule with the amount of the radioactive element that remains. One of the most commonly used age-dating measurements is the decay of uranium isotope-238 into lead-206. It takes 9 billion years for all uranium-238 to decay into lead-206. The half-life of uranium-238 is 4.5 billion years, about the age of the earth. In the process of radioactive decay a series of daughter elements are produced. They are thorium, actinium, radon, polonium, bismuth and finally lead-206. Lead-206 is not radioactive.

Decay rates of isotopes of other elements than uranium and carbon can also be used to age-date rocks. Potassium-40 decays into argon-40 in 2.5 billion years; rubidium-87 decays into strontium-87 in 48.8 billion years, and samarium-147 decays into Neodymium in 106 billion years. Carbon-14 decays to nitrogen-14 in 50,000 years.

I will focus on two techniques of radiometric age-dating: uranium-lead and carbon-nitrogen (although, the decay of potassium to argon and rubidium to strontium are frequently used to measure geologic time). All of these elements have different half-lives. Using two different radioactive elements to age-date the same sample is very strong evidence for the geologic age of associated fossils.

Uranium

Uranium begins to decay immediately after minerals containing it crystallize from magma (melted rock). Crystallization fixes the date when radioactive decay begins because the uranium incorporated in crystal lattices of minerals has a full complement of isotopes. Thereafter, radioactive decay of uranium continues until all of isotope-238 sequestered in crystal lattices of other minerals is transmuted into lead-206. Because the rate of decay is constant, the amount of lead-206 within a crystal lattice can be used to measure geologic time. Isotope chemistry can make a very accurate separation of lead-206 from the usual isotope of lead (207).

During the crystallization of magmas, the mineral zircon incorporates a few parts-per-million uranium in their crystal lattices. Zircons are highly resistant to melting during metamorphism. This makes them the most frequently used mineral to age-date very old rocks. If the temperature of metamorphism melts a rock, the age-date of that rock is reset in most minerals that contain uranium in their crystal lattices. This does not usually happen with zircons. Most of the oldest rocks in the earth's crust have been metamorphosed more than once, often separated by several hundreds of million years; nonetheless, their ancient age can be measured by measuring the amount of uranium-238 and lead-206 in the crystal lattices of zircons.

Carbon

Carbon has three isotopes—12, 13 and 14. Only carbon-14 is radioactive. Carbon-14 is continually created in the uppermost atmosphere by ultraviolet radiation reacting with molecules of carbon in carbon dioxide. Molecules of carbon dioxide are continually incorporated into tissues, shells and bones of all living organisms as long as they are alive.

Within the tissues of living organisms, carbon-14 continually decays to nitrogen-14, but it is also continually renewed. Plants continually renew carbon-14 by photosynthesis and animals continually renew carbon-14 by grazing on herbaceous plants or eating meat. After the death of an animal or plant, carbon-14 is no longer renewed and it continually decays to nitrogen-14, which is not radioactive.

Carbon-14 has a half-life of 5,700 years. Half of carbon-14 isotopes decay to nitrogen-14 in 5,700 years. Thereafter, the remaining carbon-14 continues to decay at a constant rate to nitrogen-14. Carbon-14 is an ideal element for age-dating recent organic remains because the ratio between carbon-14 and carbon-12 in dead organisms (wood, bones) dates the age of the organism at the time of its death. By 50,000 years, no measurable carbon-14 remains. When no measurable carbon-14 remains, fossils are 50,000 years old or older.

Measurements of carbon isotope-14 in the bones of extinct Pleistocene megamammals (mammoths, ground sloths, saber tooth tigers) accurately measures the time of their extinctions. All of these extinctions in Europe, Asia, Australia and North America occurred less than 50,000 years ago. These extinction dates have a very high correlation with *Homo sapiens* migration from Africa.

CREATION SCIENCE

Creation science is a contradiction. Creation science is pseudo science. It has its own assumptions that cannot be used by natural scientists to frame experiments that are replicable. Stated in different words, creation science is designed to validate the account of creation in the Genesis myth by asserting that the biosphere is so complex that it had to have had an intelligent designer. The principle technique used by creationists who claim to be scientists is to pluck examples of biologic complexity from scientific literature without understanding their origin. Thereafter, these persons use deductive reasoning to claim the truth of the Genesis myth. This is the principal purpose of creation science.

Believers in creationism share five assumptions:

1. Complexity of the biosphere demonstrates that it was designed by god.
2. God designed the biosphere for a single purpose.
3. That purpose was human benefit.
4. From the day that life began god has actively guided the operation of the biosphere.
5. Teaching evolution by natural selection actively promotes atheism because natural selection is a totally material process.

Intelligent design was presented to the public in 1984 in the book, *The Mystery of Life's Origin: Reassessing Current Theories*. It was written by Charles B. Thraxton with help from two others. All three had scientific training. Although the text is essentially secular, the authors chose to strongly emphasize biological phenomena that had an inadequate scientific explanation, especially the inability of molecular biology and related biological sciences to explain the origin of life.

Thraxton was accurate in asserting that our current understanding of natural causes does not explain the origin of life. His accompanying assertion, however, is not accurate. This assertion claims that it is impossible for natural science to explain the origin of life because there is no known molecular chemistry that can explain the spontaneous reactions of molecular chemistry that can create life. This allegation is a leap of faith that is identical with the leap of faith required to believe the god explanation for the creation of life in the Genesis myth. There is no scientific evidence that god created life, or life was created for the specific purpose of human benefit, but there is accumulating evidence that life could have originated by spontaneous reactions of molecular chemistry. As of 2009, how life originated is an open question.

The best apology for intelligent design is Phillip E. Johnson's book, *Darwin on Trial* (1991). Johnson is an old earth creationist who believes that Holy Scripture is inspired by god but is not literally true or infallible. He is a trial lawyer who has mastered the skills of argumentation before judges and juries. The jury he wants to convince is the nonscientific American public.

His argument is a plea for the operation of intelligent design from the time god created life to the present; and that intelligent design is a valid scientific concept that explains the purpose of life. The unstated assumption of his argument is that god created the biosphere for human benefit. Johnson clearly states the reason for writing the book:

> Evolution is taught in public schools...not as a theory but as a fact...Many dissidents...deny that evolution is a fact and insist that an intelligent Creator caused all living things to come into being in furtherance of a purpose...The concept of creation in itself does not imply opposition to evolution, if evolution means only a gradual process by which one kind

> of living creature changes into something different. A Creator might well have employed such a gradual process as a means of creation. Evolution contradicts creation only when it is explicitly or tacitly defined as fully naturalistic evolution—meaning evolution that is not directed by any purposeful intelligence. (P. 3–4.)

Old earth creationists concede that evolution may have taken place but they agree with young earth creationists that the complexity of the biosphere could not have been formed by random mutations preserved by natural selection. The principal think tank of fundamentalist Christians who believe in intelligent design is the Creation Research Society. Its membership is young earth creationists. Voting members are required to sign a creed that asserts:

1. Holy Scripture is the written word of god.
2. Holy Scripture is historically and scientifically accurate.
3. The account of the origin of life in Genesis is a simple historical truth.
4. All living organisms, including humans, were created by direct acts of god during the six days of creation described in Genesis.
5. The great flood that Noah survived was a worldwide historical event.

No scientific research is funded by the Creation Research Society. At a trial (1982) that contested the constitutionality of an Arkansas statute that mandated teaching the Genesis myth as an alternative to evolution by natural selection, lawyers for the scientific community asked lawyers representing fundamentalist Christians to submit articles that had been published in a scientific journal where articles receive peer review before publication.

They submitted no articles. All "research articles" funded by the Creation Research Society were published in religious journals.

The only purpose of the Creation Research Society is to use carefully selected biological examples to disprove evolution. The creation research they claim to have done is presented to audiences in a popular culture format. Almost always, "expert" speakers have an undergraduate degree in biology and an honorary PhD awarded by an unaccredited bible college.

Examples used by creationists are always separated from comparisons with research conducted by trained scientists. The vast amount of details in the scientific literature is deliberately ignored, deliberately misquoted or, more likely, not understood. In other words, their message is void of scientific integrity. Their ignorance of how scientific research is conducted or their rejection of science allows them to assert that the biology of living organisms is too complex to have evolved by natural selection.

Based on this assertion, they deduce that natural selection is a false hypothesis. There has to be another explanation because biological complexity cannot be an accident of natural selection. Biological complexity requires a designer. The designer is an omnipotent god because there is no alternative.

Believers in the Genesis myth, intelligent design and creation science reject the evidence of paleontology and geochronology that the biosphere has evolved during the four billion years that life has existed on earth. They reject evidence that the biosphere is a product of random mutations in which natural selection decides which mutations are failures and which ones are beneficial. They reject the geologic record of extinctions because fossils were buried in sediments by the flood that drowned all animals except those saved on Noah's ark. They believe that god has guided all aspects of life, and god's guidance is clearly recognizable because it is visible everywhere and at all times for those who seek it.

It is very difficult for true believers to accept that the biosphere is a product of random mutations in all visible and invisible organisms and that evolution is an ongoing process that began about four billion years ago. It is equally inconceivable to them that humans evolved from random mutations in the genes of primates because this makes humankind just another animal in the paleontological record of life on earth. They believe that humans are god's surrogate to govern the biosphere and that humankind's status at the top of the food chain is proof that god designed the biosphere for human usage. Furthermore, they believe that human management of the biosphere will continue until god ends time by rapturing true believers into heaven.

Intelligent design and its companion, creation science, is clearly a religious doctrine because it is impossible to measure omnipotence, omniscience or eternity because they have no physical properties. It is impossible for scientists to design experiments to test the Genesis account of creation because there are no definitions or assumptions that are common to creation science and natural science. Seen from the perspective of experimental science, intelligent design is a variety of science fiction.

PUBLIC POLICY

Evolution became a political issue sometime about 1910 because enough scientific evidence had accumulated to indicate that evolution was a hypothesis with a high degree of certainty. Fundamentalist Christians were increasingly concerned that its entry into school curriculums would make literate persons into godless atheists. They believed that teaching evolution by natural selection must be excluded from public school curriculums in order to prevent atheism from being taught to impressionable youths.

William Jennings Bryan became the national spokesman for American fundamentalism. He was nominated three times to be president of the United States (1896, 1900, 1908) and lost all three elections. His core constituency was agrarians who were concentrated in the plains states and cotton south. A high percentage was wheat farmers and white sharecrop cotton cultivators in ex-Confederate states. White cotton sharecrop cultivators had very high rates of illiteracy, and the isolated wheat farmers of Kansas and Nebraska had a grossly deficient understanding of how science was molding the rapidly growing urban culture of the United States and the world.

Bryan spoke for the agrarian social order. He had a profound distrust of the new urban industrial culture of the United States because he believed it was a threat to the agrarian democracy that was the foundation of American liberty. He believed that native-born white agrarians were the moral backbone of the United States, and that education curriculums being used in cities lacked the moral values that made the United States into a great nation.

During the 1920s Bryan led a national campaign to prohibit the teaching of evolution in public schools because of its atheistic assumption. His first success was Tennessee, where the legislature passed a statute in 1925 that made it "unlawful for any teacher to teach any theory that denies the Story of Divine Creation of man as taught in the Bible, and to teach instead that man has descended from a lower order of animal." The legislatures of Arkansas, Louisiana and Mississippi passed similar legislation.

In Dayton, Tennessee, a high school biology teacher, John T. Scopes, was arraigned for teaching evolution. The prosecution wanted to turn the trial into a creation versus evolution debate but soon had second thoughts. They reconsidered their strategy because the defense mobilized a formidable number of prominent scientists with national and international reputations to testify for the high probability that evolution and natural selection did, in fact, operate to create new species, including humans. The prosecution could not find any anti-evolution scientist with a reputation.

Both the prosecution and defense agreed to limit the charges against Scopes to a single issue. Did he or did he not teach evolution and thereby violate state law. Bryan accepted an invitation to testify as an expert on religion because he wanted to use the trial to convince the doubtful that the account of creation in Genesis was an actual historical event. Once in the witness box, Bryan was available for questioning by defense lawyers.

Defense lawyers asked Bryan how many events in the bible he could rationally accept as literally true. Many of his answers ascribed events to miracles performed by god. For example, Bryan had no rational explanation for how Joshua lengthened the day by making the sun stand still; nor how fish survived the flood that killed all life on earth except species that Noah saved in the ark. The result of the defense's cross examination was that Bryan was shown to be no expert on the bible, nor any scientific discipline, especially the geology and biology of evolution.

Ten years after the Scopes trial more than 70 percent of high school textbooks omitted discussion of evolution. Its teaching was not resumed until the early 1960s when funds from the national government were used to improve precollege scientific education in order to attract more students into scientific disciplines needed in the cold war and space race. The reappearance of evolution in high school textbooks generated a strong reaction from fundamentalist Christians. They wanted intelligent design to be taught in public schools as an alternative explanation (superior explanation) to evolution.

In 1965 the statutes passed by the four state legislatures in the 1920s that forbade teaching evolution, were still in force.

They were, however, rarely enforced. In 1965, the Arkansas Educational Association decided to challenge the constitutionality of the Arkansas anti-evolution statute. They claimed that biblical creationism was a religious doctrine that violated the U.S. constitution because the first amendment in the bill of rights required the separation of church and state. In practice, this meant that any public school that received any monetary support from the national government had to exclude religion from their curriculums.

A district court judge agreed with the Arkansas Educational Association and declared the statute unconstitutional. His decision was immediately appealed to the Arkansas Supreme Court where the decision was reversed. It was then appealed to the United States Supreme Court (*Epperson v Arkansas*). In 1968 the U.S. Supreme Court reversed the decision of the Arkansas Supreme Court. It was unconstitutional to teach biblical creationism in public schools because it was a religious doctrine.

The response of fundamentalist Christians to the Epperson decision was to claim that the Genesis myth could be validated by creation science. Creation science would disguise biblical creationism as a secular science and qualify it for equal time in classrooms. By 1980, bills had been introduced in twenty-seven state legislatures that mandated equal classroom time for teaching creation science and evolutionary sciences. These bills died in committee except in Arkansas and Louisiana where they were reported out of committee and voted into law in 1981.

Proponents of equal classroom time for teaching biblical creation claimed that teaching evolution by natural selection would undermine the religious convictions and moral values of students. In an effort to make creation science a valid science, a definition of creation science was included in the Arkansas statute. It was totally deductive from the Genesis myth.

1. The earth and its biosphere are the result of recent creation.
2. There has been insufficient numbers of mutations for natural selection to have evolved all of the earth's organisms from a single self-replicating cell.
3. The morphological changes of plants (vegetables in the cabbage family) and animals (breeds of dogs) are within the limits of recent creation.
4. Structural geology, geomorphology and the fossil record are due to catastrophism, including a worldwide flood.
5. There was a separate creation for humans.

The constitutionality of the Arkansas statute was immediately contested. In order to sustain the constitutionality of the statute the state had to argue that creation science was as valid as evolutionary science. This required expert witnesses. The state had great difficulty finding reputable scientists (those who had actually conducted experiments) to testify that creation science was science. The state could not put the director of the Institute of Creation Science in the witness box because the tenor of all of his publications was an apology for fundamentalist religion. The principal evidence in his publications was revelation. He would have experienced the same probing questions that exposed William Jennings Bryan as ignorant of experimental science at the Scopes trial in 1925.

The United States Supreme Court declared the Arkansas statute unconstitutional because creation science was not science. It was a sectarian religious doctrine that wanted to use classrooms for religious proselytizing. In 1987 the Louisiana statute was also ruled unconstitutional by the Court in *Edwards v Aguillard*. The Court's opinion said that the Louisiana statute "advances a religious doctrine by requiring either the banishment of the theory of evolution from public school classrooms or the presentation of a religious viewpoint that rejects evolution entirely."

Fundamentalist Christians were forced to revise their strategy in order to prevent evolution from being taught in public schools.

They have used several strategies to exclude evolution from high school curriculums: (1) lobby state legislatures to pass statutes that forbid teaching evolution in public schools, (2) gain control of state and local school boards in order to require teachers to omit teaching evolution, (3) adopt textbooks that omitted a discussion of evolution, (4) establish private schools where the Genesis myth is taught as a true historical event.

SUMMARY

During the past five million years several species of great apes evolved bipedalism, but only one great ape evolved a giant brain. A giant brain is more efficient than big bodies, big teeth or high speed to reach the top of the food chain. The evolution of a giant brain makes humans the equivalent of *Tyrannosaurus rex* in all niches they choose to contest.

Belief in the Genesis myth that claims that god has chosen humans to occupy the top of the food chain is an artifact of having a giant brain that can generate abstract concepts. These concepts can be religious mythology (Genesis myth) or scientific hypotheses (evolution by natural selection) to explain humanity's place in the biosphere. The Genesis myth originated as revelation. The assertion that it is a real event allows believers to deduce that an omnipotent god designed the biosphere for the benefit of humans. No science is required to validate this claim because it is an assertion based on revelation. In contrast, evolution by natural selection is based on observation and its validity is sustained by experiments that can be replicated. These experiments indicate that evolution is a hypothesis with a high degree of certainty.

What should people believe? A myth based on revelation or observations that are confirmed by experiments that can be replicated? Additionally, experiments in many other life sciences confirm that natural selection is a hypothesis that has a high degree of certainty.

All disciplines of natural science indicate that humankind's tenure at the top of the food chain is an extremely recent event. It began approximately 50,000 years ago when humans invented many new varieties of edged stone tools. Until these inventions, humans were just another large mammal competing for survival in the East African highlands. Possession of these tools propelled a worldwide migration. Will humans remain at the top of the food chain? The geologic record clearly indicates that the tenure of all large animals at the top of the food chain is very tenuous and often very short. There is a high probability that extinction will be the fate of humankind.

CHAPTER 6

HUMAN EXTINCTION

Will humans become another animal in the ongoing mass extinction of megamammals? Can humans become like mammoths, mastodons, ground sloths, glyptodonts, saber tooth cats, aurochs, Irish elk and Neanderthal humans that became extinct during the last 25,000 years? Is it possible that humans will join rhinoceros, hippopotamus, gorillas, orangutans, tigers, leopards, buffalo (bison), muskoxen and many species of antelope that are on the edge of extinction? The answer is yes.

How will human extinction happen? Why will it happen? When will it occur? Will it be due to one cause or a confluence of many causes? These four questions can be summarized by one question: Can warfare cause the extinction of humankind? The answer is yes. I will analyze two global scenarios of human extinction: (1) suicidal warfare, (2) end of time warfare that will precede the rapture of true believing Christians into heaven and consign all sinners to hell.

SUICIDAL WAR

Nuclear warfare is suicidal warfare, and nuclear warfare in conjunction with germ warfare will produce obliterating casualties. Some contemporary regional wars and some future regional wars have a potential for exploding into a global nuclear catastrophe that has been fictionalized by a number of people, including Nevil Shute in *On the Beach*. A global nuclear war is warfare in another dimension than warfare with conventional weaponry because there will be little distinction between military and civilian deaths. The enemy includes all persons who can be mobilized to produce any product or service needed to sustain warfare. The battlefield will be everywhere.

ENDTIME WAR

The endtime war is the apocalypse. True believing Christians believe that the endtime war will be between the armies of the good god named Lord and the armies of the evil god named Satan. The army of the Lord will triumph because Christ will return to earth to lead the army of god to victory. Unbelievers will be converted or go to hell. After the triumph, Christ will gather all true believers (living and dead) and they will bodily ascend into heaven (the rapture). They will have eternal life in perfect peace and joy.

The final battle of the apocalypse will be fought at Armageddon. Armageddon is an Old Testament battle site located in the nation of Israel where a victory by a Jewish army preserved the Jews and their religion from conquest by an evil empire. The apocalypse will precede the global triumph of Christianity and end the separation between earth and heaven. Time will end, and so will human occupancy on earth.

Is the endtime war a fantasy? Yes. The scenario of an apocalypse and the subsequent rapture of true believers is certainly a religious delusion. It is science fiction dressed in religious robes that exists only in the minds of true believers. The scenario of a global nuclear war is also fantasy—until righteous political leaders on earth make it happen. Will it happen? Possibly. Righteous political leaders, who consider themselves chosen by god, may use nuclear weapons in conjunction with germ warfare to exterminate cultures that form an axis of evil. Such action could ignite a global war of extermination. Both fantasies will produce the same result—human extinction.

FUNDAMENTALISM

The term fundamental became current after 1910. The name was coined by Southern Baptists but rapidly included Pentecostals and Churches of God. It defined Christian beliefs that were beyond compromise. The principal fundamental beliefs are the following: (1) literal truth of the bible, (2) creationism, (3) being born again to ensure salvation, (4) the apocalypse will occur in near time, (5) second coming of Christ will happen in near time, (6) rapture for true believers. As noted previously, these doctrines had their roots early in the 1800s among persons who believed that the democratic uniqueness of the United States required that the United States evangelize the world to establish democracy and Christianity.

Certitude of salvation is the great attraction of fundamentalist Christianity. The experience of being born again by being baptized by the Holy Spirit creates a direct relationship between an individual and god. Certitude of salvation is instantly available to all persons who are born again. Neither an institutional church nor the eucharist (a Christian sacrament) is essential for salvation. Salvation is obtained within a local congregation. Born again Christians know they are saved because they have accepted Jesus as a personal savior. It is a certitude that is similar to the certitude that Roman Catholics seek by full participation in the sacraments of the church and obeying the teachings of the papacy.

Most American fundamentalists share five other beliefs: (1) prohibition of abortion; (2) practicing faith healing; (3) hostility for using public funds for scientific research, except medicine, agriculture and armaments, because most hypotheses that frame scientific inquiries are godless; (4) distrust of the global economy because there is no certitude in how it operates; (5) uncritical support for the political agenda of Israel because Jerusalem is god's city and the final battle of the apocalypse will be fought at Armageddon in Israel.

TRUE BELIEF

True belief requires certitude. The certitude sought by true believers is salvation. Fundamentalist clergy teach that there is only one source of certitude. It is their interpretation of the bible. They believe that all of its words are literally true.

Fundamentalist clergy are trained to exhort sinners that salvation is available to all persons who repent their sins and accept Jesus as their personal savior. This assures them of salvation. How do they know they are saved? They experience baptism by the Holy Spirit. They are born again. This experience usually takes place in an emotionally charged church service that may produce a trance among the newly saved individuals. After recovery, the congregation knows that the soul of a person who has been born again will enjoy eternal life if their future conduct is exemplary.

SALVATION

Salvation is repentance of sins by the power of faith. Eternal life for a person's soul is god's reward to people who forgive those who sin against them. The consistent message of true believing clergy is that all contemporary cultures are drowning in sin. Repentant sinners who have experienced baptism by the Holy Spirit become as free from sin as newborn children. Being born again is the ticket required for the salvation of a person's soul and for bodily ascension into heaven (rapture) if Christ's second coming is in their lifetime.

> For the Lord himself will descend from heaven with a cry of command, with the archangel's call, and with the sound of the trumpet of God. And the dead in Christ will rise first; then we who are alive, who are left, shall be caught up together with them in the clouds to meet the Lord in the air; and so we shall always be with the Lord. Therefore comfort one another with these words. But as to the times and the seasons, brethren, you have no need to have anything written to you. For you yourselves know well that the day of the Lord will come like a thief in the night. (1 Thessalonians 4:16–18; 5:1–2)

The rapture will occur after seven years of war (tribulation), followed by a climatic battle (apocalypse) between the armies of the Lord and the armies of Satan. The Lord's army will triumph in a battle at Armageddon and the rapture will follow. The rapture will end time because all persons who reject Jesus as their savior will be dead or in hell and all born again Christians will be in heaven.

The rapture is the salvation myth of true believers. It is the equivalent of the creation myth in Genesis, but instead of explaining why god created life on earth and chose humans to govern the biosphere, it explains why god will end human life on earth and end human governance of the biosphere. In the Genesis myth humans are the special creation of god to do his work on earth. In the rapture myth god completes the purpose of humans on earth. Time will end when true believers are raptured into heaven and evil persons are consigned to hell. Humans will be extinct on earth.

BIBLICAL PROPHECY

Biblical prophecy is interpreting selective texts in the bible that validate the fantasies of the apocalypse and rapture. Clergy claim that these concepts are fundamental Christian beliefs. Biblical prophecy has no rules. It can be practiced by any clergy who wants to invent the future. Fundamentalist clergy have a free hand to select any biblical text that they claim will prove that contemporary violent events were predicted in the Old and New Testaments.

The best texts are in the Old Testament book of Daniel and the New Testament book of Revelation. Fantasy in these books is used to deduce that the end of time is near and that repentance is necessary to be included in the rapture when it comes. Alternatively, if sinners do not repent they will be consigned to hell when time ends.

This was not why these books were written and why bishops in the early church selected them for inclusion in the bible. The purpose of these books was to encourage Jews and Christians to keep the faith during times of persecution. The book of Daniel was written about 150 BC and the book of Revelation was written about AD 80. Both books are full of violence, but it is the violence of fantasy rather than the violence of historical events. This is in contrast to the many accounts of real violence recorded in the books of Samuel, Kings and Chronicles.

The fantasy dreams in Daniel and Revelation are allegories that depict god directing his wrath on governments persecuting believing Jews and Christians. The message of these books is that these governments will disappear but the faithful will survive. Until god acts, the faithful must suffer and endure. The reward for their faithfulness is salvation and eternal life.

Fundamentalist clergy who preach biblical prophecy interpret contemporary violent events as the prelude to the seven years of tribulation before the apocalypse. But the apocalypse never comes. When violent events are resolved, as the end of World War II or the dissolution of communist governments in Europe—and the apocalypse does not happen—it is because these stressful events were insufficiently violent for god to initiate the time of tribulation. Fundamentalist clergy continually replace failed predictions of the impending apocalypse by new stressful events. There is a never-ending supply of them in the real world. This scenario cannot be wrong because deductive reasoning, based on fantasy texts in an inerrant bible, makes tribulation and the apocalypse into facts of future history.

Is the world now experiencing the seven years of tribulation preceding the apocalypse? Almost all fundamentalist clergy say yes because the new stressful event is Muslim terrorism that is focused on the nation of Israel and this terrorism radiates outward into the rest of the world, especially the United States. They know with certainty that Muslim terrorism portends the apocalypse, but the exact future date is unknown. In the meantime, they must wait and prepare; and the best preparation is being born again to ensure salvation when it comes.

DEDUCTIVE REASONING

Michael D. Evans is one of the principal practitioners of deductive reasoning that foretells the coming of the time of tribulation. In *American Prophecies* (254), he states, "The season of Christ's return is definitely near." In the following quotes Evans tells his readers how he arrived at this conclusion. "The prophet Daniel accurately predicted the kingdoms and governments that would follow ancient Babylon's disappearance from the world scene in Daniel 2:31–45. Many of Daniel's prophecies have already been fulfilled, and his writings constitute the cornerstone of biblical prophecy" (*American Prophecies*, 252).

Let me quote Daniel 2:31–45 so that readers may judge how fundamentalist clergy deduce that the apocalypse is near.

You saw, O king, and behold a great image. This image, mighty and of exceeding brightness, stood before you, and its appearance was frightening. The head of this image was of fine gold, its breast and arms of silver, its belly and thighs of bronze, its legs of iron, its feet partly of iron and partly of clay. As you looked, a stone was cut out by no human hand, and it smote the image on its feet of iron and clay, and broke them in pieces; then the iron, the clay, the bronze, the silver, and gold, all together were broken in pieces and became like the chaff of the summer threshing floors; and the wind carried them away, so that not a trace of them could be found. But the stone that struck the image became a great mountain and filled the whole earth.

This was the dream; now we will tell the king its interpretation. You, O king, the king of kings, to whom the God of heaven has given the kingdom, the power, and the might, and the glory, and into whose hand he has given, wherever they dwell, the sons of men, the beasts of the field, and the birds of the air, making you rule over them all—you are the head of gold. After you shall arise another kingdom inferior to you, and yet a third kingdom of bronze, which shall rule over all the earth. And there shall be a fourth kingdom, strong as iron, because iron breaks to pieces and shatters all things; and like iron which crushes, it shall break and crush all these. And as you saw the feet and toes partly of potter's clay and partly of iron, it shall be a divided kingdom; but some of the firmness of iron shall be in it, just as you saw iron mixed with miry clay. And as the toes of the feet were partly iron and partly clay, so the kingdom shall be partly strong and partly brittle. As you saw the iron mixed with miry clay, so they will mix with one another in marriage, but they will not hold together, just as iron does not mix with clay. And in the days of those kings the God of heaven will set up a kingdom which shall never be destroyed, nor shall its sovereignty be left to another people. It shall break in pieces all these kingdoms and bring them to an end, and it

> shall stand for ever; just as you saw that a stone was cut from a mountain by no human hand, and that it broke in pieces the iron, the bronze, the clay, the silver, and the gold. A great God made known to the king what shall be hereafter. The dream is certain, and its interpretation sure.

It should be clear to most persons who read this text that fundamentalist clergy can deduce any fantasy they want from it and similar texts in Daniel and the book of Revelation. These texts are licenses for fundamentalist clergy to interpret any contemporary stressful event as part of the wars of tribulation that will precede Christ's second coming.

For true believers, the concept of an endtime war is wholly rational because they do not believe in geologic time. Evolution did not happen. Time began 8,000 to 10,000 years ago, according to the Genesis myth, and time may end tomorrow, next year or ten years from now. The concept of the second coming of Christ and the end of time is a concept that true believers can deduce from many texts in an inerrant bible, but especially the books of Daniel and Revelation.

The apocalypse and rapture are the intellectual property of fundamentalist Christians. They continually think about the apocalypse because their clergy continually bombard them with the duty to be ready for the second coming of Christ. As endtime approaches, fundamentalist clergy urge Christians to assure the salvation of their souls while they are still alive by being baptized by the Holy Spirit and accepting Jesus Christ as their personal savior. Assurance is obtainable by a public show of repentance and a public experience of being born again. An educated clergy is not necessary for preaching the redeeming power of the Holy Spirit.

SCIENCE

The preaching credentials for most contemporary fundamentalist clergy are two years of training at an unaccredited bible college where no science is taught. Fundamentalist clergy are indifferent to (or discourage) higher education in secular universities because exposure to science is dangerous for faith. Fundamentalist Christians distrust science because it teaches critical skills that challenge (or ignore) the certitudes of faith embodied in revealed religion.

Foremost among the critical skills taught at secular universities is inductive reasoning because inductive reasoning governs the conduct of all scientific research. They especially reject the conclusions of the life sciences of zoology, genetics, molecular biology and paleontology that hypothesizes that the purpose of life is reproduction and humans are just another animal in the biosphere that random mutations put at the top of the food chain. They reject the interpretation that the fundamental purpose of life is reproduction, even though there is a high degree of certainty that reproduction has been continuous for about four billion years. God is irrelevant to this hypothesis because the survival of life (and its diversification) is dependent on natural selection, not creationism.

Fundamentalist Christians distrust scientific research because it cannot verify their principal spiritual concerns—true faith and the certitude of salvation. The response of most scientists to these claims is: salvation from what? Salvation is irrelevant to science because its dimensions cannot be physically measured. If salvation exists, the only evidence is subjective faith. Salvation that is based on faith is not part of science where all hypotheses are open to continuous modification by experiments that use physical measurements and inductive reasoning.

The attempt to impose deductive reasoning on science comes

into clear focus when fundamentalist clergy use deductive reasoning to validate the pseudoscience of intelligent design deduced from the Genesis myth. Fundamentalist rejection of evolution and supporting biological sciences is a repeat of the papacy's rejection of Galileo's evidence that the earth is one of several planets that orbit the sun (1633)—and that the earth is not the center of the universe.

There is a third distrust of science that receives less attention. It is the age of the universe. Recent telescopic measurements estimate its age to be 14 billion years. If, however, creation was only 10,000 years ago, 14 billion years is a fantasy number for true believers. They ignore this number and concentrate on defending the validity of the Genesis myth.

True believing Christians are certain that human life has a higher purpose than reproduction because the Genesis myth claims that humans are a special creation of god. Salvation is more important than any scientific hypothesis because god's spokesmen (prophets) have revealed that salvation and eternal life can be obtained on earth by faith and repentance. This certitude is recorded in the bible and is available to all persons who seek it.

WAR

War is a form of predation waged by humans against humans. Human history records many wars of extermination. These wars were usually conducted by societies with overwhelming technical superiority or overwhelming numbers. The most common purpose of wars of extermination is to gain control of land in order to increase the subsistence opportunities of the dominant people. Other motivations for conventional war are eliminating threats to national security, dynastic rivalries, imposing religious conformity, forcibly appropriating existing wealth, gaining commercial advantages or enslaving humans to perform labor for the benefit of governing elites. Almost all persons in almost all cultures understand these motivations and approve or oppose them.

Nuclear wars will not kill all humans, but aftermath events could cause extinction. Three of these events are famine, plague and anarchy. It is unlikely that these three events will occur simultaneously or be of sufficient magnitude to cause human extinction. But, if there were a series of closely spaced nuclear wars, in timespans of ten, twenty or thirty years, the global population would be reduced after each war. These wars could push humanity into a slide toward extinction.

Who are the people who believe that nuclear warfare could cause human extinction? Secular persons understand that nuclear warfare has the potential for causing human extinction. Mostly, however, they do not think about it. When they do think about it, they understand that exterminating warfare is likely to occur if righteous political leaders use nuclear weapons to obliterate evil empires.

FAMINE

A nuclear war could destroy the complex commercial and political relationships that are essential for the operation of industrial cultures. Food production is especially vulnerable to disruption. Food to feed industrial cities depends on intensive cultivation. The inputs necessary for intensive cultivation depend on industrial inputs and a transportation network that can carry fertilizers to farmers and food to cities.

If fertilizer factories, transportation networks and sources of fuel were destroyed, urban food supplies would plummet.

Nuclear obliteration of cities would have a cascading effect on food production. Manufactured inputs into agriculture would cease and the ability to transport food to urban markets would be seriously impaired or would cease. Surviving cities could not be fed. Regional famines would become predictable events.

PLAGUE

The breakdown of sanitary facilities (sewers, potable water, preventive medicines like vaccinations), plus reduced nutrition, would increase possibilities for plagues. The best example of a destabilizing plague was the arrival of European settlers in the Americas in the centuries following Columbus's voyages. Military conquests are the stuff of history books—as they should be—because they are dramatic events; however, plagues are less dramatic but are more potent in reducing human populations.

All indigenous societies in the Americas were decimated by diseases brought by European conquerors, traders, cultivators and African slaves. The great killer was smallpox, but measles, tuberculosis and other contagious diseases contributed their share of deaths. Most plague deaths occurred within fifty years of the initial conquests as European trade and settlement extended inland. Plagues, combined with advanced technologies of war, extinguished these cultures overnight.

Plagues are usually consigned to footnotes even though they made it possible for small numbers of Europeans to displace all indigenous cultures, including those that had dense populations at the time of conquest. How many people died in these plagues? This depends on the estimated populations of indigenous cultures. Population estimates vary from A to Z. It is, however, known with a high degree of certainty that population densities plummeted after the arrival of Europeans, and that these declines were due to plagues of contagious diseases. The larger the affected populations, the greater number of plague deaths.

Cultural anthropologists disagree about the percentage of indigenous populations that succumbed to plague deaths, but whatever the numbers, plague deaths were very high in the years following the arrival of Europeans. The best records are from agricultural cultures that practiced intensive cultivation. These cultures attracted Europeans who kept written records from the day of their arrival. These cultures were concentrated in Mesoamerica (Mexico) and the Andes highlands (Peru).

What was the death rate in societies practicing intensive cultivation? Again, estimates of death rates are highly variable. They are, however, more closely bunched because there are references to population declines in documents written by European observers. This gives their estimates some degree of credibility. The estimated percentages for indigenous cultures in Mexico and Peru varied from 80 to 95 percent, but these cultures had sufficient numbers so that enough of the population developed antibodies for a population base to survive.

In sparsely populated hunter–gatherer societies, high population losses would have a different effect. A 90 percent loss of population in these societies equals extinction. Surviving persons would be forced to join other cultures that had large enough populations to survive catastrophic losses. Frequently, this meant becoming voluntary slaves.

Spanish conquistadors did not want catastrophic population declines because they wanted to substitute themselves for dead or displaced indigenous aristocracies. They wanted to retain the indigenous corvee labor systems that had built pyramids in Mexico and Yucatan and paved pathways on mountainous terrain in Peru. They would use the corvee labor of peasants for different projects. They would mine gold and silver ores and build cart and wagon roads to tidewater on the Gulf of Mexico and the Pacific coast of Peru in order to expedite marine communications with Europe.

On the islands of Cuba, Hispaniola and Puerto Rico, corvee labor would cultivate sugar cane and tobacco for sale in Europe. The catastrophic decline in the indigenous population on these islands forced the Spanish to import African slaves to perform the labor to cultivate these crops. Africans had the requisite antibodies to resist high death rates from the plagues that decimated indigenous populations.

In contrast, the nations that colonized North America welcomed plagues because they removed an obstacle to inland expansion. Inland expansion in North America was not based on seeking indigenous societies that had precious metals to loot or ores to mine; although, Europeans searched for them. Inland expansion in North America was propelled by a search for agricultural land and the exploitation of other resources, especially forest products. Settlers wanted to displace indigenous populations that blocked access to these resources. Plagues had the desired effect.

Indigenous societies north of the Rio Grande were thinned by severe to catastrophic population losses. Thereafter, everywhere that Europeans contested control of land use, minimal military power by colonial governments was able to displace them.

Western Europe experienced similar population losses by the Black Death (bubonic plague) that entered Europe in 1346 and ran its course to 1352. In those years somewhere between 25 and 33 percent of the population perished with population losses as high as 70 percent in some cities. Urban population losses were higher than village losses because public sanitation was primitive to nonexistent. Inadequate sewage disposal supported large rat populations that were the vectors for the fleas that carried the pathogens of the bubonic plague.

ANARCHY

The historical record has many examples where famine, war and anarchy were of sufficient magnitude to cause depopulation and cultural collapse. One of the best examples is the collapse of Mayan civilization that centered on the Yucatan peninsula of Mexico. The evidence for collapse is clearly visible.

Many temple complexes of monumental size were abandoned and became overgrown by tropical forests. At the time they were built, they were located in the middle of intensively cultivated land that supported dense populations. The populations that built them disappeared before the arrival of Europeans. Currently, the temple complexes at Tikal, Palenque, Bonampak, Copan and elsewhere are in process of being disinterred from tropical forests because they have become tourist attractions.

Peasants began settling the Yucatan peninsula about 2,300 years ago. By AD 250 increasing population densities had induced the practice of intensive cultivation. Sustaining intensive cultivation required some variety of central governance. In Yucatan this governance was supplied by ministates with ruling aristocracies that were highly efficient at mobilizing corvee labor to build and maintain an agricultural infrastructure, monumental temples and other buildings for regal and festival uses. Accompanying the organization of numerous ministates was an increasing frequency of internecine warfare.

By AD 800 population densities were beginning to create serious problems in food production because the soil on the limestone karst that forms most of the Yucatan peninsula is of marginal fertility. Infertile soil was aggravated by lack of surface sources of water. Thereafter, there was almost continuous internecine warfare among the ministates. Mayan civilization collapsed soon after AD 950.

The great question is: what caused the collapse of Mayan

culture? The principal cause was probably virulent internecine warfare. The food and manpower demands for war triggered peasant rebellions against ruling aristocracies. Peasants killed the aristocracy. When a peasant rebellion succeeded in one ministate it had a domino effect on neighboring ministates. The death of governing aristocracies ended food taxation, corvee labor to build and maintain temples and the mobilization of warriors for internecine warfare.

The end of aristocratic governance had a catastrophic effect on agriculture and social cohesion. Anarchy and famine ensued. The labor required to maintain the infrastructure of intensive cultivation (irrigation, reservoirs, terraces, raised fields) was not done. A series of drought years further intensified food stress. Famine conditions recurred with increasing frequency. Depopulation accelerated. Archeology, biology, limnology, carbon-14 age-dating and Mayan texts provide substantial evidence that this chain of events extinguished Mayan civilization.

The best evidence of the Mayan collapse comes from deciphering the Mayan script. A group of persons in one ministate borrowed a script from a Mexican society and improved it. It had four uses in both Mayan and Mexican cultures: (1) devising and transmitting a calendar that accurately fixed dates for planting maize, (2) recording the amount of food taxes levied on peasant villages, (3) recording the number of men who could be mobilized from each villages for warfare or for building temples, (4) recording dynastic accomplishments in warfare and building temples.

Most surviving examples of the Mayan script are chiseled in stele (stone monuments) that record victories in war, temple completions, dynastic successions and calendars. After about AD 950 no stele were erected at any Mayan temple complex because central governance ceased. Peasant populations crashed.

By AD 1250 huge areas of land surrounding temple complexes were devoid of human habitations. In large areas of the Yucatan peninsula population declines were more than 90 percent.

EXTINCTION

Will a global nuclear war, followed by anarchy, famine and plagues be the apocalypse? This is a possibility. The cumulative effects of prolonged nuclear warfare are easy to catalog, but would they be sufficient to destroy advanced industrial cultures? Probably.

The following is a list of events that would continue to reduce populations after a prolonged nuclear war has drastically reduced populations worldwide:

1. Weakened or destroyed central governments.
2. Contraction or disappearance of intensive agriculture and greatly reduced food production.
3. Famines.
4. Disappearance of public health services provided by central governments.
5. Surviving population would be highly susceptible to cholera and dysentery vectored by unsanitary water.
6. Production of medicines and vaccines would cease and contagious diseases like typhus, tuberculosis, mumps, measles and diphtheria would greatly increase death rates in famine-weakened populations.
7. Child mortality would soar to sixteenth century rates.
8. AIDS would be an instant plague.
9. The largest killer would probably be genetically engineered pathogenic viruses used as an adjunct weapon during nuclear warfare.

10. Almost certainly, famine and plagues would generate conditions of anarchy.
11. Surviving populations would retribalize. Internecine warfare would invariably follow in order to gain control of land for subsistence cultivation or to enslave prisoners to cultivate land to produce food to feed warriors.
12. The slide to extinction could begin.

It is highly likely that surviving humans would revert to barbarism by using surviving weaponry to conduct internecine warfare. Is revision to barbarism the same as extinction? No, but a catastrophic decline of populations into fragmented tribes creates conditions for these societies to serially disappear by famines, plagues and internecine warfare, or a concurrence of these causes with other unpredictable stresses.

AFTER HUMAN EXTINCTION

There are no known exceptions to two rules in the geologic record: (1) mega-animals at the top of the food chain are the first animals to become extinct whenever catastrophic stresses impact the biosphere, (2) life continues after mass extinctions.

Will the posthuman earth be a Garden of Eden? No. The biosphere will revert to the pre-human biosphere where natural selection operates without human interference. Organisms with generalized reproductive capacities will temporarily fill most vacant niches. Natural selection will continue to operate but it will be a genetic free-for-all. During the two million years following human extinction, many new organisms will evolve to fill specialized niches. This will be especially true for vacant niches of mega-animals.

SUMMARY

What is the probability of the occurrence of a biblical apocalypse? Zero! What is the probability of human extinction by nuclear and germ warfare? It is possible. The ability of humans to wage suicidal wars makes it clearly possible for humankind's tenure at the top of the food chain to be of short duration.

What becomes of god after human extinction? Without humans on earth, the concept of god ceases to exist. God will be extinct.

Eras of Geologic Time

MILLIONS OF YEARS	GEOLOGIC ERAS
Present	**mass extinction**
65–present	Tertiary (Cenozoic)
65	**mass extinction**
142–65	Cretaceous
206–142	Jurassic
206	**mass extinction**
251–206	Triassic
251	**mass extinction**
290–251	Permian
362–290	Carboniferous (Mississippian, Pennsylvanian)
362	**mass extinction**
417–362	Devonian
438–417	Silurian
435	Plant and animals colonize land
438	**mass extinction**
495–438	Ordovician
495	**mass extinction**
550–495	Cambrian
610–550	Vendian
555–550	Ediacaran Assemblage
610	First metazoan animals
700	First metazoan organisms
1500	First eukaria cells
3500	First fossil bacteria
4000	Approximate beginning of life
4600	Beginning of planet earth

Bibliography

Allen, Keith C., Derek E. G. Briggs, eds., *Evolution and the Fossil Record*, Washington, Smithsonian Institution Press, 1990.

Alvarez, Walter, *T. Rex and the Crater of Doom*, Princeton, Princeton University Press, 1997.

Belcher, Stephen, *African Myths of Origin*, London, Penguin Books, 2005.

Bengtson, Stefan, E*arly Life on Earth*, New York, Columbia University Press, 1994.

Benton, Michael, *When Life Nearly Died: The Greatest Mass Extinction of All Time*, London, Thames and Hudson, 2003.

Brack, Andre, ed., *The Molecular Origins of Life: Assembling Pieces of the Puzzle*, Cambridge, Cambridge University Press, 1998.

Bryson, Bill, *A Short History of Nearly Everything*, New York, Broadway Books, 2003.

Carroll, Sean B., *Remarkable Creatures: Epic Adventures in the Search for the Origin of Species*, Boston, Houghton Mifflin Harcourt, 2009.

Chouard, Tanguy, "Beneath the Surface," *Nature*, Vol. 456, November 2008, 300–303.

Cohen, Claudine, *The Fate of the Mammoth: Fossils, Myth, and History*, Chicago, University of Chicago Press, 2002.

Cowan, Richard, *History of Life*, Boston, Blackwell Scientific Publications, 1990.

Coyne, Jerry A., *Why Evolution is True*, New York, Viking, 2009.

Crick, Francis, *Of Molecules and Men*, Amherst, Prometheus Books, 2004 (reprint of 1966 edition).

Cross, Whitney A. *The Burned-Over District*, New York, Harper Torchbooks,1965

Dawkins, Richard, *River Out of Eden*, New York, Basic Books, 1995.

Dawkins, Richard, *The Ancestor's Tale: A Pilgrimage to the Dawn of Evolution*, New York, Houghton Mifflin, 2004.

Day, Michael H., *Guide to Fossil Man*, Chicago, University of Chicago Press, 1986.

Deamer, David W., Elizabeth H. Mahon, Giovanni Bosco, "Self-Assembly and Function of Primitive Membrane Structures," in Stefan Bengtson, ed. *Early Life on Earth*, New York, Columbia University Press, 1994.

Deamer, David W., "Origins of Membrane Structure," in Lynn Margulis, Clifford Matthews, Aaron Haselton, eds. *Environmental Evolution: Effects of the Origin and Evolution of Life on Planet Earth*, Cambridge, MIT Press, 2000.

Deamer, David W., "How Leaky Were Primitive Cells?" *Nature*, Vol. 454, July 2008.

Diamond, Jared M., *Guns, Germs, and Steel: The Fates of Human Societies*, New York, Norton, 1999.

Diamond, Jared M., *Collapse: How Societies Choose to Fail or Succeed*, Penguin Books, New York, 2005.

Erwin, Douglas H., *Extinction: How Life on Earth Nearly Ended 250 Million Years Ago,* Princeton, Princeton University Press, 2006.

Evans, Michael D., *The American Prophecies: Ancient Scriptures Reveal Our Nation's Future*, New York, Warner Faith, 2004.

Fedonkin, Mikhail A., "Vendian Faunas and the Early Evolution of Metazoa," in Jere H. Lipps, Philip W. Signor, eds. *Origin and Evolution of the Metazoa*, New York, Plenum Press, 1992.

Fraser, Nicholas, *Dawn of the Dinosaurs: Life in the Triassic*, Bloomington, University of Indiana Press, 2006.

Gould, Stephen Jay, *Ever Since Darwin: Reflection in Natural History*, New York, Norton, 1979.

Gould, Stephen Jay, *Wonderful Life: The Burgess Shale and the Nature of History*, New York, Norton, 1989.

Grieve, Richard A., Eugene M. Shoemaker, "The Record of Past Impacts on Earth," in Tom Gehrels, ed. *Hazards due to Comets and Asteroids*, Tucson, University of Arizona Press, 1994.

Hagee, John, *Jerusalem Countdown: A Warning to the World*, Lake Mary, Florida, Front Line Publishers, 2006.

Hallam, Anthony, ed., *Patterns of Evolution as Illustrated by the Fossil Record*, Amsterdam, Elsevier Scientific Publishing, 1977.

Hou, Xian-Guang, Richard J. Aldridge, Jan Bergstrom, David J Siveter, Derik J. Siveter, Feng Xiang-Hong, *The Cambrian Fossils of Chengjiang, China: The Flowering of Early Animal Life*, Blackwell Publishing, 2007.

Ings, Simon, "An Eye for the Eye," *Nature*, Vol. 456, November 2008, 304–309.

Johnson, Phillip E., *Darwin on Trial*, Washington, Regnery Gateway, 1991.

Kirsch, Jonathan, *A History of the End of the World: How the Most Controversial Book in the Bible Changed the Course of Western Civilization*, New York, HarperSanFrancisco, 2006.

Kirschner, Marc W., John C. Gerhart, *The Plausibility of Life: Resolving Darwin's Dilemma*, New Haven, Yale University Press, 2005.

Knoll, Andrew H., *Life on a Young Planet: The First Three Billion Years of Evolution on Earth*, Princeton, Princeton University Press, 2003.

Lange, Ian M., *Ice Age Mammals of North America: A Guide to the Big, the Hairy, and the Bizarre*, Missoula, Mountain Press, 2002.

Laporte, Leo F. ed., *Evolution and the Fossil Record*, San Francisco, Freeman, 1978.

Lazcano, Antonio, "The Transition from Nonliving to Living," in Stefan Bengtson, ed., *Early Life on Earth*, New York, Columbia University Press, 1994.

Lipps, Jere H., Philip W. Signor, eds., *Origin and Early Evolution of the Metazoa*, New York, Plenum Press, 1992.

Lurquin, Paul F., *The Origins of Life and the Universe*, New York, Columbia University Press, 2003.

Mann, Charles C., *New Revelations of the Americas before Columbus*, New York, Vintage Books, 2005.

Mansy, Sheref S., Jason P Schrum, Mathangi Krishnamurthy, Sylvia Tobe, Douglas A. Treco, Jack W. Szostak, "Template- directed Synthesis of a Genetic Polymer in a Model Protocell," *Nature,* Vol. 454, July 3, 2008.

Margulis, Lynn, *Symbiosis in Cell Evolution: Life and Its Environment on the Early Earth*, San Francisco, Freeman, 1981.

Margulis, Lynn, Clifford Matthews, Aaron Hasselton, eds. *Environmental Evolution: Effects of the Origin and Evolution of Life on Planet Earth*, Cambridge, MIT Press, 2000.

Margulis, Lynn, Michael F. Dolan, *Early Life: Evolution on the Precambrian Earth*, Boston, Jones and Bartlett Publishers, 2002.

Moore, John A., *From Genesis to Genetics: The Case of Evolution and Creationism*, Berkeley, University of California Press, 2002.

Morris, Simon Conway, *The Crucible of Creation: The Burgess Shale and the Rise of Animals*, Oxford, Oxford University Press, 1998.

Oro, Juan, "Early Chemical State in the Origin of Life," in Stefan Bengtson, ed., *Early Life on Earth*, New York, Columbia University Press, 1994.

Phillips, Kevin, *American Theocracy: The Perils and Politics of Radical Religion, Oil, and Borrowed Money in the 21st Century*, New York, Viking, 2006.

Pierson, Berverly K., "The Emergence, Diversification, and Role of Photosynthetic Eubacteria," in Stefan Bengtson, ed., *Early Life on Earth*, New York, Columbia University Press, 1994.

Powner, Matthew W., Beatirce Garland, John D. Sutherland, "Synthesis of Activated Pyrimidine Ribonucleotides in Prebiotically Plausible Conditions," *Nature*, Vol 459, May 14, 2009, 171-172, 239-242.

Quammen, David, *The Song of the Dodo: Island Biogeography in the Age of Extinctions*, New York, Scribner, 1996.

Rampino, Michael R., Bruce M. Haggerty, "Extraterrestrial Impacts and Mass Extinctions of Life," in Tom Gehrels, ed., *Hazards due to Comets and Asteroids*, Tucson, University of Arizona Press, 1994.

Ranger, Terence O., Paul Slack, eds., *Epidemics and Ideas: Essays on the Historical Perception of Pestilence*, Cambridge, Cambridge University Press, 1992.

Rose, Kenneth D., *The Beginning of the Age of Mammals*, Baltimore, Johns Hopkins University Press, 2006.

Runnegar, Bruce, "Proterozoic Eukaryotes: Evidence from Biology and Geology," in Stefan Bengtson, ed., *Early Life on Earth*, New York, Columbia University Press, 1994.

Schopf, J. William, "The Oldest Known Records of Life: Early Archean Stromatolites, Microfossils, and Organic Matter," in Stefan Bengtson, ed., *Early Life on Earth*, New York, Columbia University Press, 1994.

Schopf, J. William, ed., *Life's Origin: The Beginnings of Biological Evolution*, Berkeley, University of California Press, 2002.

Scott, Eugenie C., *Evolution vs Creationism: An Introduction*, Westport, Greenwood Press, 2004.

Seavoy, Ronald E., *Famine in Peasant Societies*, Westport, Greenwood Press, 1986.

Selden, Paul A., John R. Nudds, *Evolution of Fossil Ecosystems*, Chicago, University of Chicago Press, 2004.

Shanks, Niall, *God, The Devil, and Darwin: A Critique of the Intelligent Design Theory,* New York, Oxford University Press, 2004.

Simonetta, Alberto M., Simon Conway Morris, eds., *The Early Evolution of Metazoa and the Significance of Problematic Taxa*, Cambridge, Cambridge University Press, 1991.

Smit, Jan, "Extinction at the Cretaceous-Tertiary Boundary: The Link to the Chicxulub Impact," in Tom Gehrels, ed., *Hazards due to Comets and Asteroids*, Tucson, University of Arizona Press, 1994.

Southwood, Richard, *The Story of Life*, Oxford, Oxford University Press, 2003.

Spong, John S., *Rescuing the Bible from Fundamentalism: A Bishop Rethinks the Meaning of Scripture*, New York, Harper San Francisco, 1992.

Steadman, David W., *Extinction and Biogeography of Tropical Pacific Islands*, Chicago, University of Chicago Press, 2006.

Stenger, Victor J., *God: The Failed Hypothesis: How Science Shows that God does not Exist*, Amherst, New York, Prometheus Books, 2007.

Stetter, Karl O., "The Lesson of Archaebacteria," in Stefan Bengtson, ed., *Early Life on Earth*, New York, Columbia University Press, 1994.

Stinchcomb, Bruce L., *World's Oldest Fossils*, Atglen, Pa., Schiffer Publishing, 2007.

Strandberg, Todd, Terry James, *Are You Rapture Ready? Signs, Prophesies, Warnings, Threats, and Suspicious that the Endtime is Now*, New York, Dutton, 2003.

Tattersall, Ian, Jeffrey H. Schwartz, *Extinct Humans*, Boulder, Westview Press, 2001.

Thraxton, Charles B., Walter L. Bradley, Roger L. Olsen, *The Mystery of Life's Origin: Reassessing Current Theories*, Escondido, California, Philosophical Library, 1984.

Wachtershauser, Gunter, "Vitalysts and Virulysts: A Theory of Self-Explaning Reproduction," in Stafan Bengtson, ed., *Early Life on Earth*, New York, Columbia University Press, 1994.

Wang, K., H. H. J. Geldsetzer, B. D. E. Chatterton, "A Late Devonian Extraterrestrial Impact and Extinction in Eastern Gondwana: Geochemical, Sedimentological, and Faunal

Evidence," in Burkhard O. Dressler, ed., *Large Meteorite Impacts and Planetary Evolution*, Boulder, Geological Society of America, Special Paper 293, 1994.

Wilson, Edward O., *The Future of Life,* New York, Vintage Books 2003.

Index

D

E

F

G

H

I

J

K

L

M

N

O

P

R

About the Author

Ronald E. Seavoy is professor emeritus of history at Bowling Green State University, Bowling Green, Ohio. Previous books by the author are *Origins of the American Business Corporation* (Greenwood, 1982); *Famine in Peasant Societies* (Greenwood, 1986); *Famine in East Africa: Food Production and Food Policies* (Greenwood, 1989); *The American Peasantry: Southern Agricultural Labor and Its Legacy*, 1850–1995 (Greenwood, 1998); *Subsistence and Economic Development* (Praeger, 2000); *A New Exploration of the Canadian Arctic* (Hancock House, 2002); *Origins and Growth of the Global Economy From the Fifteenth Century Onward* (Praeger, 2003); *An Economic History of the United States: From 1607 to the Present* (Routledge, 2006).